T0292395

Heterogeneous Reconfigurable Processors for Real-Time Baseband Processing

Chenxin Zhang • Liang Liu • Viktor Öwall

Heterogeneous Reconfigurable Processors for Real-Time Baseband Processing

From Algorithm to Architecture

 Springer

Chenxin Zhang
Department of Electrical and Information
    Technology
Lund University
Lund, Sweden

Liang Liu
Department of Electrical and Information
    Technology
Lund University
Lund, Sweden

Viktor Öwall
Department of Electrical and Information
    Technology
Lund University
Lund, Sweden

ISBN 978-3-319-24002-2   ISBN 978-3-319-24004-6   (eBook)
DOI 10.1007/978-3-319-24004-6

Library of Congress Control Number: 2015958762

Springer Cham Heidelberg New York Dordrecht London

Printed on acid-free paper

Springer International Publishing AG Switzerland is part of Springer Science+Business Media (www.springer.com)

Contents

Contents

List of Acronyms

ADC	Analog-to-digital converter.
ALU	Arithmetic logic unit.
ASIC	Application-specific integrated circuit.
ASIP	Application-specific instruction set processor.
AWGN	Additive white Gaussian noise.
BPSK	Binary phase-shift keying.
CFO	Carrier frequency offset.
CGRA	Coarse-grained reconfigurable architecture.
CISC	Complex instruction set computing.
CMAC	Complex-valued multiply-accumulate.
CMOS	Complementary metal-oxide-semiconductor.
CORDIC	Coordinate rotation digital computer.
CP	Cyclic prefix.
CSI	Channel state information.
CVG	Candidate vector generation.
DAC	Digital-to-analog converter.
DFE	Digital front-end.
DLP	Data-level parallelism.
DMA	Direct memory access.
DSP	Digital signal processor.
DVB	Digital video broadcasting.
DVB-H	Digital video broadcasting for handheld.

ED   Euclidean distance.
EPA   Extended pedestrian A.
EQD   Equally distributed.
ETU   Extended typical urban.
EVA   Extended vehicular A.

FEC   Forward error correction.
FER   Frame error rate.
FFT   Fast Fourier transform.
FIFO   First in first out.
FNE   Fast node enumeration.
FPGA   Field-programmable gate array.
FSD   Fixed-complexity sphere decoder.
FSM   Finite-state machine.
FU   Function unit.

GALS   Globally asynchronous locally synchronous.
GOPS   Giga operations per second.
GPC   Generic signal processing cell.
GPP   General purpose processor.
GPR   General purpose register.
GPS   Global positioning system.
GPU   Graphics processing unit.
GSM   Global system for mobile communications.

HDL   Hardware description language.

i.i.d.   Independent and identically distributed.
ICI   Inter-carrier-interference.
IFFT   Inverse fast Fourier transform.
ILC   Inner loop controller.
ILP   Instruction-level parallelism.
IMD   IMbalanced distributed.
IP   Intellectual property.
ISA   Instruction set architecture.
ISI   Inter-symbol-interference.

LS   Least square.
LSB   Least significant bit.
LTE   Long term evolution.

LTE-A   Long term evolution-advanced.
LUT   Look-up table.

MAC   Multiply-accumulate.
MGS   Modified Gram-Schmidt.
MIMO   Multiple-input multiple-output.
ML   Maximum-likelihood.
MMR   Matrix mask register.
MMSE   Minimum mean-square error.
MPMC   Multi-port memory controller.
MRC   Maximum-ratio combining.
MSE   Mean squared error.

NFC   Near field communication.
NoC   Network-on-chip.
NRE   Non-recurring engineering.

OFDM   Orthogonal frequency division multiplexing.

PDP   Power-delay profile.

QAM   Quadrature amplitude modulation.
QoS   Quality of service.
QPSK   Quadrature phase-shift keying.
QRD   QR decomposition.

RAM   Random access memory.
RC   Resource cell.
RISC   Reduced instruction set computing.
ROM   Read-only memory.
RTL   Register transfer level.

SCC   Stream configuration controller.
SCENIC   SystemC environment with interactive control.
SD   Sphere decoder.
SDC   Stream data controller.
SDR   Software-defined radio.
SIMD   Single instruction multiple data.

SIMT	Single instruction stream multiple tasks.
SNR	Signal-to-noise ratio.
SPE	Successive partial node expansion.
SQRD	Sorted QR decomposition.
STS	Short training symbol.
SVD	Singular value decomposition.
TDD	Time-division duplexing.
TLP	Thread-level parallelism.
UART	Universal asynchronous receiver/transmitter.
UMTS	Universal mobile telecommunications system.
VDP	Vector dot product.
VHDL	Very high speed integrated circuit (VHSIC) HDL.
VLIW	Very long instruction word.
VLSI	Very-large-scale integration.
VPR	Vector permutation register.
WCDMA	Wideband code division multiple access.
ZF	Zero-forcing.

List of Definitions

$(\cdot)^*$	Complex conjugate.
$(\cdot)^H$	Hermitian transpose.
$(\cdot)^T$	Vector/matrix transpose.
$(\cdot)^\dagger$	Matrix pseudo-inverse.
$(\cdot)_i$	Column vector.
$(\cdot)_{i,i}$	$(i, i)^{\text{th}}$ matrix element.
$\| \cdot \|$	Euclidean vector length.
$\| \cdot \|_2$	ℓ^2-norm.
M	Constellation size.
$N_{\mathbf{SW}}$	Frequency correlation window in R.MMSE-SW.
N_c	Number of OFDM subcarriers.
N	Number of antennas.
$\mathbf{H}$	Complex-valued MIMO channel matrix.
$\mathbf{I}$	Identity matrix.
$\mathbf{P}$	Permutation matrix in sorted-QR decomposition.
$\mathbf{Q}$	Unitary matrix in QR decomposition.
$\mathbf{R}$	Upper triangular matrix in QR decomposition.
$\mathbf{n}$	i.i.d. complex Gaussian noise vector.
$\lfloor \cdot \rfloor$	Floor function. Rounds x to nearest integer towards $-\infty$.
$\mathcal{O}$	Computational complexity.
σ_n^2	Variance of noise vector $\mathbf{n}$.
Δf	OFDM subcarrier spacing.
s_{CMOS}	Technology scaling factor.
$\approx$	Approximation.
$\lceil \cdot \rceil$	Ceiling function. Rounds x to nearest integer towards $+\infty$.

$\odot$ Element-wise vector multiplication.
ε Fractional carrier frequency offset.
η Post-detection SNR.

$\in$ For $x \in A$, the element x belongs to the set A.

$\boldsymbol{\Omega}$ Node perturbation parameter in MMSE-NP.

$\propto$ Proportional.

$\mathbb{R}$ Real continuous space.

$\mathcal{Q}$ Slicing function in symbol detection, returning a nearest constellation point.

θ OFDM symbol start.

Chapter 1
Introduction

This book discusses an interdisciplinary study in wireless communication and Very-large-scale integration (VLSI) design, more specifically, implementation of digital baseband processing using reconfigurable architectures. Development of such kind of systems, sometimes referred to as baseband processors [15] or Software-defined radio (SDR) platforms [8], is an important and challenging subject, especially for small-scale base stations (e.g., femtocells) and mobile terminals that must provide reliable services under various operating scenarios with low power consumption.

The importance of the subject is driven by two facts. First, there is a huge demand for wireless communication in the world. The number of devices connected to the Internet in one way or the other is expected to reach 50 billion by 2020 [3, 9]. In other words, every person on earth will have around six devices on average. Second, the number of radio standards grows increasingly fast in order to suffice ever-growing user demands such as data rate. For example, compared to the world's first hand-held device demonstrated in 1973, today's fourth-generation (4G) mobile terminals are able to process not only voice and text but also data streaming with the speed of up to gigabit-per-second [7]; the coming 5G wireless communication network will provide 1000-fold gain in capacity. Moreover, modern wireless systems need to be backward compatible to support 2G Global system for mobile communications (GSM) and 3G Universal mobile telecommunications system (UMTS), as well as to support a range of different radio standards for improving user experience. Examples of these standards are bluetooth, IEEE 802.11 series, Global positioning system (GPS), and Near field communication (NFC). As envisioned in [5], a single 4G mobile terminal needs to support more than 10 radio standards with tens of operation modes in each standard [e.g., 63 for 3GPP Long term evolution (LTE)]. Using traditional implementation strategies, equipping each of these standards with an Application-specific integrated circuit (ASIC), becomes antiquated and unaffordable with regard to area consumption and development time.

© Springer International Publishing Switzerland 2016
C. Zhang et al., *Heterogeneous Reconfigurable Processors for Real-Time Baseband Processing*, DOI 10.1007/978-3-319-24004-6_1

Besides, it is unlikely that a user will enable all of these standards at the same time in a single terminal. Thus, there is a need for a flexible hardware platform capable of supporting operations among multiple standards and tasks and allocating resources dynamically to suffice current computational demands.

In addition to the multi-standard multi-task support, flexibility is required to cope with the rapid evolution of baseband processing algorithms and enable run-time algorithm adaption to provide better Quality of service (QoS) and maintain robust, reliable, and seamless connectivity. Furthermore, benefiting from the hardware reconfigurability, such architectures have the potential to perform system updates and bug-fixes while the system is in operation. This feature will prolong product life-time and ensure benefits in terms of time-to-market [13, 15, 16]. Last but not the least, from an algorithm development perspective, reconfigurable computing provides a more software-centric programming approach. This allows hardware platforms to be developed on-demand and potentially in the same language as used for software development. Unified programming environment enhances productivity by simplifying system integration and verification. Besides its importance, the target subject faces many design challenges in practical implementations, such as requirements of high computational performance and low energy consumption. Primary concerns for contemporary system designs are shifting from computational performance to energy efficiency [2, 17]. This trend becomes more and more prominent in wireless communication designs. For example, the transition from 3G to 4G wireless communication systems demands 3 orders of magnitude increase in computational complexity, whereas the total power budget remains approximately constant in a single mobile terminal [14, 21]. Reconfigurable architectures, since its invention in 1960 [10], promise to offer great hardware flexibility and computational performance. They allow run-time hardware reconfigurations to accelerate arbitrary algorithms, and thus extend the application domain and versatility of the device. However, due to huge routing overhead, they cannot match power and area efficiency of ASICs, in spite of their tremendous developments over the past decades. As an example, fine-grained interconnects in commercial Field-programmable gate array (FPGA) consume over 75 % of the chip area [20], and cause 17–54 times area overhead and 5.7–62 times more power consumption in comparison to ASICs [12]. Moreover, bit-level function blocks of FPGAs incur additional area and power penalties when implementing word-level computations. The area and power overhead have restricted the usage of reconfigurable architectures in cost-sensitive applications such as wireless communication in mobile terminals. To address these overhead issues, new types of reconfigurable architectures with coarse-grained function blocks have gained increasing attention in recent years in both academia and industry [1, 4, 6, 11, 18, 19].

This book presents a coarse-grained dynamically reconfigurable cell array architecture, which is designed and tailored with a primary focus on digital baseband processing in wireless communication. By exploiting the computational characteristics of the target application domain, the presented domain-specific cell array architecture bridges the gap between ASICs and conventional reconfigurable platforms. The flexibility, performance, and hardware efficiency of the cell array are demonstrated through case studies.

1.1 Scope of the Book

The goal of this book is to find efficient reconfigurable architectures that can provide a balance among computational capability, flexibility, and hardware efficiency. The driving application for hardware developments and performance evaluations is digital baseband processing in wireless communication. The target platform is commercially deployed wireless communication equipment and devices, which need to provide real-time performance with restricted budgets of physical size and energy dissipation.

The central part of this book is the presentation of a dynamically reconfigurable cell array architecture. Performance of the cell array is evaluated through two case studies, which are conducted to address two following questions:

- Can the cell array be used for multi-standard and multi-task processing? Is the control overhead affordable?
- Can the cell array meet real-time requirements when performing sophisticated baseband processing tasks? Under such a use case, what is the area and energy efficiency in comparison to ASICs and conventional reconfigurable architectures?

Throughout the book and by conducting algorithm–architecture co-design, special attention is paid to four distinct areas of the cell array design:

- System architecture design, including various processing elements, memory sub-systems, Network-on-chip (NoC), and dynamic reconfiguration.
- Design flow of the cell array.
- Design trade-offs, including selection of processing elements and accelerators, task partitioning between hardware and software as well as between processing elements and memory sub-systems.
- Instruction set and function descriptor design for various processing elements and memory sub-systems, respectively.

Digital baseband processing in wireless communication systems includes many tasks such as Orthogonal frequency division multiplexing (OFDM) modulation/demodulation, Multiple-input multiple-output (MIMO) signal processing, Forward error correction (FEC), interleaving, scrambling, etc. Among these, this book focuses on four crucial blocks in a typical baseband processing chain at the receiver, i.e., Digital front-end (DFE), channel estimation, channel pre-processing, and symbol detection. However, the same design methodology is applicable for other baseband processing blocks and applications.

1.2   Outline

Chapters 2 and 3 serve to give an overview of the research field. Chapter 2 discusses
reconfigurable architectures and various processing alternatives. Chapter 3 covers
typical digital baseband processing tasks in contemporary wireless communication
systems. These two introductory chapters are not intended to give detailed descrip-
tions on each of the subject. They are presented to give reference information on
terms and concepts used later in the book.

Chapter 4 introduces the coarse-grained dynamically reconfigurable cell array
architecture, including both system infrastructure and a hardware design flow. Using
the cell array as a baseline architecture, Chaps. 5 and 6 present two case studies to
demonstrate the performance of the presented domain-specific reconfigurable cell
array. The two studies are conducted in accordance to the processing flow of a
typical baseband processing chain at the receiver. In addition, the two case studies
manifest architectural evolution of the cell array, namely from scalar- to vector-
based architecture. Chapter 7 opens up discussion on reconfigurable architecture
design for next-generation wireless communication systems. Signal processing
operations in Massive MIMO and design challenges for reconfigurable platforms
are discussed.

1.2.1   Chapter 4: The Reconfigurable Cell Array

Conventional fine-grained architectures, such as FPGAs, provide great flexibility
by allowing bit-level manipulations in system designs. However, the fine-grained
configurability results in long configuration time and poor area and power efficiency,
and thus restricts the usage of such architectures in time-critical and area/power-
limited applications. To address these issues, recent work focuses on coarse-grained
architectures, aiming to provide a balance between flexibility and hardware efficiency by adopting word-level data processing. In this chapter, a coarse-grained
dynamically reconfigurable cell array architecture is presented. The architecture
is constructed from an array of heterogeneous functional units communicating
via hierarchical network interconnects. The strength of the architecture lies in the
simplified data sharing achieved by decoupled processing and memory cells, the
substantial communication cost reduction obtained by a hierarchical network structure, and the fast context switching enabled by a unique run-time reconfiguration
mechanism.

1.2.2 Chapter 5: Multi-Standard Digital Front-End Processing

This chapter aims at demonstrating the flexibility of the reconfigurable cell array architecture and evaluating the control overhead of hardware reconfigurations, in terms of clock cycles and area consumption. For this purpose, the cell array is configured to concurrently process multiple radio standards. Flexibility of the architecture is demonstrated by performing time synchronization and Carrier frequency offset (CFO) estimation in a digital front-end receiver for multiple OFDM-based standards. As a proof-of-concept, this book focuses on three contemporarily widely used radio standards, 3GPP LTEs, IEEE 802.11n, and Digital video broadcasting for handheld (DVB-H). The employed reconfigurable cell array, containing 2×2 resource cells, supports all three standards and is capable of processing two concurrent data streams. Dynamic configuration of the cell array enables run-time switching between different standards and allows adoption of different algorithms on the same platform. Thanks to the adopted fast configuration scheme, context switching between different operation scenarios requires at most 11 clock cycles.

1.2.3 Chapter 6: Multi-Task MIMO Signal Processing

This chapter aims at demonstrating the flexibility and real-time processing capability of the cell array as well as evaluating the area and energy efficiency when performing sophisticated baseband processing tasks.

Driven by the requirement of multi-dimensional computing in contemporary wireless communication technologies, reconfigurable platforms have come to the era of vector-based architectures. In this chapter, the reconfigurable cell array is extended with extensive vector computing capabilities, aiming for high-throughput baseband processing in MIMO-OFDM systems. Besides the heterogeneous and hierarchical resource deployments, a vector-enhanced Single instruction multiple data (SIMD) structure and various memory access schemes are employed. These architectural enhancements are designed to suffice stringent computational requirements while retaining high flexibility and hardware efficiency. To demonstrate its performance and flexibility, three computationally intensive blocks, namely channel estimation, channel pre-processing, and symbol detection, of a 4×4 MIMO processing chain in a 20 MHz 64-QAM 3GPP Long term evolution-advanced (LTE-A) downlink are mapped and processed in real-time.

1.2.4 Chapter 7: Future Multi-User MIMO Systems

This chapter looks ahead into advanced multi-user Massive MIMO technology for 5G wireless communication systems and opens up discussion for its baseband

processor design. Wireless communication technology is evolving at a fast pace
to meet requirements of emerging applications. Accordingly, the developed reconfigurable architecture should be extensible to support signal processing in future
wireless communication systems. In this chapter, the basic concept of the relatively
new MIMO technology, Massive MIMO, is introduced. To facilitate the corresponding hardware architecture design, operations in Massive MIMO baseband
processing are profiled and analyzed. Additionally, we discuss how the new features
in Massive MIMO processing affect the architecture design, in terms of operation
characteristics and processing distribution. This chapter serves as a pre-study and a
design guideline for developing an efficient reconfigurable computing platform for
Massive MIMO systems.

References

1. Z. Abdin, B. Svensson, Evolution in architectures and programming methodologies of coarse-grained reconfigurable computing. Microprocessors Microsyst. Embed. Hardw. Des. **33**, 161–178 (2009)
2. S. Borkar, Thousand core chips - a technology perspective, in *44th Annual Design Automation Conference (DAC)*, 2007, pp. 746–749
3. Broadcom, Facts at a glance, Apr 2014. https://www.broadcom.com/docs/company/BroadcomQuickFacts.pdf
4. A. Chattopadhyay, Ingredients of adaptability: a survey of reconfigurable processors, in *VLSI Design*, Jan 2013
5. F. Clermidy, et al., A 477mW NoC-based digital baseband for MIMO 4G SDR. in *IEEE International Solid-State Circuits Conference (ISSCC)*, Feb 2010, pp. 278–279
6. K. Compton, S. Hauck, Reconfigurable computing: a survey of systems and software. ACM Comput. Surv. **34**, 171–210 (2002)
7. E. Dahlman, S. Parkvall, J. Skold, *4G: LTE/LTE-Advanced for Mobile Broadband*, 1st edn. (Academic, New York, 2011)
8. M. Dillinger, K. Madani, N. Alonistioti, *Software Defined Radio: Architectures, Systems and Functions*, 1st edn. (Wiley, New York, 2003)
9. Ericsson, White paper: more than 50 billion connected devices - taking connected devices to mass market and profitability, Feb 2011. http://www.akos-rs.si/files/Telekomunikacije/Digitalna_agenda/Internetni_protokol_Ipv6/More-than-50-billion-connected-devices.pdf
10. G. Estrin, Organization of computer systems: the fixed plus variable structure computer, in *Western Joint IRE-AIEE-ACM Computer Conference*, May 1960, pp. 33–40
11. R. Hartenstein, A decade of reconfigurable computing: a visionary retrospective, in *Design, Automation Test in Europe Conference Exhibition (DATE)*, 2001, pp. 642–649
12. I. Kuon, R. Tessier, J. Rose, FPGa architecture: survey and challenges. Found. Trends Electron. Des. Autom. **2**(2), 135–253 (2008)
13. T. Lenart, Design of reconfigurable hardware architectures for real-time applications. Ph.D. thesis, Department of Electrical and Information Technology, Lund University, May 2008
14. G. Miao, N. Himayat, Y. Li, A. Swami, Cross-layer optimization for energy-efficient wireless communications: a survey. Wirel. Commun. Mob. Comput. **9**(4), 529–542 (2009)
15. A. Nilsson, Design of programmable multi-standard baseband processors. Ph.D. thesis, Department of Electrical Engineering, Linköping University, 2007
16. H. Svensson, Reconfigurable architectures for embedded systems. Ph.D. thesis, Department of Electrical and Information Technology, Lund University, Oct 2008

17. M.B. Taylor, A landscape of the new dark silicon design regime. IEEE Micro **33**(5), 8–19 (2013)
18. R. Tessier, W. Burleson, Reconfigurable computing for digital signal processing: a survey. J. VLSI Signal Process. Syst. **28**, 7–27 (2001)
19. T.J. Todman, G.A. Constantinides, S.J.E. Wilton, O. Mencer, W. Luk, P.Y.K. Cheung, Reconfigurable computing: architectures and design methods. Comput. Digit. Tech. **152**, 193–207 (2005)
20. C.C. Wang, F.L. Yuan, H. Chen, D. Marković, A 1.1 GOPS/mW FPGA chip with hierarchical interconnect fabric, in *IEEE Symposium on VLSI Circuits (VLSIC)*, June 2011, pp. 136–137
21. M. Woh, S. Mahlke, T. Mudge, C. Chakrabarti, Mobile supercomputers for the next-generation cell phone. IEEE Comput. **43**(1), 81–85 (2010)

Chapter 2
Digital Hardware Platforms

Since the invention of the integrated circuit in the 1950s, there has been explosive developments of electronic circuits. Over the last decades, the amount of transistors, which are the fundamental elements of digital and analog circuits, fitting on a single silicon die has increased exponentially, from a few thousands to billions to date. This trend was already observed in 1965 [15] by Intel's co-founder Gordon E. Moore and later came to be known as "Moore's law" coined by Carver Mead. Moore's law has held true since then and is a driving force of the advancements of Very-large-scale integration (VLSI) design [11].

Enabled by the technology advancements, various forms of hardware platforms emerged to cater to a variety of applications. Depending on design trade-offs between flexibility and efficiency, these platforms can be broadly divided into three classes, namely *programmable processors*, *reconfigurable architectures*, and *Application-specific integrated circuits (ASICs)*. Programmable processors include, for example, General purpose processors (GPPs) and Application-specific instruction set processors (ASIPs). Reconfigurable architectures differ from the programmable processors in a way that they expose both data and control path to the user and are "programmable" through hardware configurations. Field-programmable gate array (FPGA) is a well-recognized example of this architecture category. ASICs are customized designs with limited flexibility. Hardware modifications after chip fabrication for new function adoption is barely possible for this type of platforms. They are commonly used in time- and power-critical systems, where flexibility is not a primary concern. Figure 2.1 illustrates a general view of how these three classes of platforms fare in the flexibility-efficiency design space. It should be pointed out that comparison of particular architecture instances among these classes has become increasingly obscure because of huge architecture varieties and different optimization objectives such as application domains and speed grades. Thus, Fig. 2.1 only serves to give an overview of how different platforms trade flexibility for efficiency. Flexibility, including programmability and

© Springer International Publishing Switzerland 2016
C. Zhang et al., *Heterogeneous Reconfigurable Processors for Real-Time Baseband Processing*, DOI 10.1007/978-3-319-24004-6_2

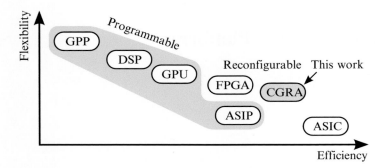

Fig. 2.1 Comparison of flexibility and efficiency for various forms of hardware platforms. This book focuses on the design of Coarse-grained reconfigurable architectures (CGRAs)

versatility, is measured as the ability to adopt a platform into different application domains and to perform different tasks. For instance, GPPs are highly flexible platforms since they are designed without having any particular application in mind. Efficiency relates to both computational performance and energy consumption and is a measure of how well a platform performs in an application. In this context, ASICs reveal the highest efficiency because of hardware customizations. This book focuses on Coarse-grained reconfigurable architectures (CGRAs), aiming to bridge the flexibility-efficiency gap between ASICs and the other two classes of platforms, illustrated in Fig. 2.1.

2.1 Programmable Processors

Programmable processors are designed based on instruction sets, which are specifications of operation codes (opcodes) used to conduct operations of underlying hardware elements. Depending on design objectives, an instruction set can be optimized with respect to, for example, application domain and friendliness to high-level programming constructs [10]. Some examples of Instruction set architecture (ISA) categories are Complex instruction set computing (CISC), Reduced instruction set computing (RISC), and Very long instruction word (VLIW).

Based upon the retargetability of the instruction set, programmable processors can be categorized into fixed and configurable ISAs. Compared to the latter one, fixed ISAs are easy to design and can be optimized for obtaining high performance such as high clock frequency by deep pipelining [10]. Examples of fixed ISAs are GPPs, special-purpose processors, and ASIPs. Configurable ISAs provide the user flexibilities in selecting appropriate instructions for target applications. This way, the ISAs can be customized to attain higher efficiency in comparison to fixed ISAs. However, this instruction set customizability complicates the design of baseline architecture and software tool chain (e.g., compiler and emulator).

2.1.1 General-Purpose Processors

GPPs are highly programmable, capable of supporting any algorithm that can be compiled to a computer program. Thus, they are dominantly used in personal computers. Although GPPs have always been implemented with the latest semiconductor technology in order to achieve the highest possible processing speed, they suffer from a performance bottleneck: the sequential nature of program execution. To address this issue, many design techniques have been proposed, which range from ISA to microarchitecture design with a goal of increasing the number of executed instructions per second. Examples of these techniques are superscalar and VLIW architectures for exploiting *Instruction-level parallelism (ILP)*, Single instruction multiple data (SIMD) architectures (e.g., Intel's Pentium MMX and AMD's 3DNow! ISA) for enabling *Data-level parallelism (DLP)*, and multithreading technology (e.g., Intel's hyper-threading [14]) for providing *Thread-level parallelism (TLP)*. Furthermore, GPPs have shifted to a multi-core paradigm due to energy and power constraints on growth in computing performance [7]. Figure 2.2 shows the slowdown in processor performance growth, clock speed, and power consumption, as well as the continued exponential growth in the number of transistors per chip [7].

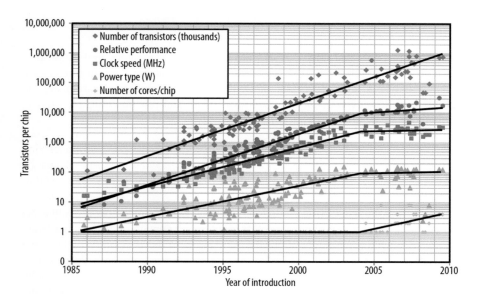

Fig. 2.2 Transistors, frequency, power, performance, and processor cores over time [7]

2.1.2   Special-Purpose Processors

Special-purpose processors are designed to be used for a particular application domain. Well-known examples are Digital signal processors (DSPs) and Graphics processing units (GPUs). DSPs are designed for performing digital signal processing tasks such as filtering and transforms. Commonly used operations in signal processing algorithms are accelerated in DSPs. An example is multiplication followed by accumulation, widely used in digital filters [18]. This operation is performed using dedicated Multiply-accumulate (MAC) units in DSPs and usually takes one clock cycle to execute. Other commonly used operations include various addressing modes such as modulo and ring-buffer.

GPUs are specialized computational units dedicated to manipulating computer graphics. Thanks to their highly parallel structure (e.g., containing hundreds of processing cores [16]), they are able to process large blocks of data in parallel. Taking advantage of the high processing capability, General-Purpose computing on Graphics Processing Unit (GPGPU) has recently gained in popularity. An example is the CUDA platform [8] from Nvidia, which supports C/C++ and Fortran programming on GPUs and can also be used for Matlab program accelerations [17].

2.1.3   Application-Specific Instruction Set Processors

Compared to DSPs and GPUs, ASIPs are optimized for a single application or a small groups of applications [13]. A general design flow is that a baseline processor, which could be a RISC processor or DSP, is extended with application-specific instructions. Besides, infrequently used instructions and function units are pruned, aiming to trade flexibility for energy and cost efficiency.

2.1.4   Configurable Instruction Set Processors

Different from the fixed ISAs, configurable instruction set processors provide users a collection of instructions and a baseline architecture containing various hardware features. Depending on target applications, users have the possibility of selecting appropriate instructions to construct a customized instruction set at design-time. Meanwhile, the microarchitecture of the processors can be customized by selecting, for example, different function units and the number of pipeline stages. Once the instruction set and the microarchitecture are fine tuned, hardware implementation of the processor is generated. From the hardware's point of view, the generated processor is a type of ASIP, however, with on-demand function customizations. Xtensa configurable cores [3] from Cadence (previously Tensilica) is an example of the configurable instruction set processor. Thanks to the instruction set and microarchitecture customizations, this type of processors provides high processing performance and hardware efficiency. However, design of the baseline architecture and the corresponding software support are more complicated than fixed ISAs, since they need to cover a huge set of configurations.

2.2 Application-Specific Integrated Circuits

ASICs are designed to perform specific tasks. Therefore, computational data paths and control circuits can be optimized for particular use cases. This brings ASICs to the far right of the design space in Fig. 2.1, indicating that they are the most efficient (in terms of performance and energy consumption) type of platforms among the three classes. Therefore, ASICs are commonly used to achieve real-time performance within the budget for physical size and energy dissipation. However, the specialized hardware architecture limits the capability of adapting system to different applications and operation scenarios. This limitation results in reduced overall area efficiency in terms of hardware reuse and sharing. Additionally, this type of platforms requires a rather long hardware redesign time (for bug-fixes or function updates) and exhaustive testing procedures. Furthermore, the exploding silicon design cost limits the adoption of ASICs, especially in deep sub-micro semiconductor technology.

2.3 Reconfigurable Architectures

Reconfigurable architectures are the ones having the capability of making substantial changes to the data path itself in addition to the control flow. This means that not only the software that runs on a platform is modified, but also how the hardware architecture operates [1, 5, 6, 9, 21, 22]. With combined control and data path manipulations, reconfigurable architectures are able to exploit potential parallelism, enable energy efficient computing, allow extensive hardware reuse, and reduce system design cycle and cost [12].

Reconfigurable architectures are either *homogeneous* or *heterogeneous*. In a homogeneous architecture, all elements contain the same hardware resources. This uniform structure simplifies the mapping of user applications, since additional constraints on function partitions and placements are avoided. However, homogeneous structures are inefficient in terms of hardware utilization of logic and routing resources [12]. In contrast, heterogeneous architectures contain array elements with different functionality, such as specialized elements for stream data processing or control-flow handling. Compared to the homogeneous structure, adoption of various types of array elements reduces hardware overhead and improves power efficiency at the cost of more complex mapping algorithms.

The size of the hardware elements inside a reconfigurable architecture is referred to as *granularity*. Fine-grained architectures and CGRAs are two variants of reconfigurable architectures. Fine-grained architectures, such as FPGAs, are usually built up on small Look-up tables (LUTs). Such architectures have the ability to map any logic functions at bit-level onto their fine-grained lattice. However, this bit-oriented architecture results in a large amount of control and routing overhead, for example, when performing word-level computations. These overheads also affect power consumption and system configuration time. In contrast, CGRAs are constructed from larger building blocks in a size ranging from Arithmetic logic units (ALUs) to full-scale processors. These hardware blocks communicate through a word-level routing network. The increased granularity in CGRAs reduces routing area overhead, improves configuration time, and achieves higher power efficiency despite less mapping flexibility. Besides, CGRAs differ from fine-grained architectures in design methodology. To map functionality into gates, FPGA designs rely on a hardware-centric approach, which usually requires programming in Hardware description language (HDL) such as VHDL. In contrast, CGRAs provide a more software-centric programming approach to map functionality to, for example, processing cores using a higher level language like C. Software-centric design approach enhances productivity and simplifies system integration and verification.

This book focuses on the development of CGRA, more specifically, domain-specific CGRA for baseband processing in wireless communication systems. Detailed architecture of the presented CGRA-based dynamically reconfigurable cell array is presented in Chap. 4 with case studies in Chaps. 5 and 6.

2.4 A Comment on Power Efficiency

As mentioned in Chap. 1, primary concerns for contemporary system designs are shifting from computational performance to power efficiency [2, 20]. Attaining high power efficiency is especially important for the target applications of this book, namely small-scale base stations and mobile terminals, since they are all constrained by stringent power requirements. Thus, it is crucial to have a better understanding of the composition of power consumption.

The total power consumption for a digital circuit built with CMOS transistors may be expressed as [19]

$$P_{\text{total}} \sim \underbrace{\alpha \cdot (C_{\text{L}} + C_{\text{SC}}) \cdot V_{\text{DD}}^2 \cdot f}_{P_{\text{dynamic}}} + \underbrace{(I_{\text{DC}} + I_{\text{leak}}) \cdot V_{\text{DD}}}_{P_{\text{leakage}}}, \qquad (2.1)$$

where P_{total}, P_{dynamic}, and P_{leakage} represent the total, dynamic, and leakage power consumption, respectively. α is the switching activity of the circuit, C_{L} the load capacitance, C_{SC} the short circuit capacitance, V_{DD} the supply voltage, and f the clock frequency. I_{DC} and I_{leak} denote the static and leakage current, respectively.

In the design of reconfigurable architectures, $P_{dynamic}$ is usually a dominating factor because of high clock frequency and hardware utilization. In comparison, the leakage power is of less concern for such kind of architectures. However, it should be pointed out that leakage power is becoming more and more important with technology scaling and thus needs more attention. One of the well-known approaches for designing low power circuits is to reduce the quadratic term V_{DD}^2 in (2.1) at the cost of performance sacrifice such as clock frequency. To compensate for the performance loss, different techniques can be used such as pipelining and parallel processing [4] but at the expense of area consumption. Thus, it can be seen that hardware designing is a trade-off between various parameters among the design space.

References

1. Z. Abdin, B. Svensson, Evolution in architectures and programming methodologies of coarse-grained reconfigurable computing. Microprocessors Microsyst. Embed. Hardw. Des. **33**, 161–178 (2009)
2. S. Borkar, Thousand core chips - a technology perspective, in *44th Annual Design Automation Conference (DAC)*, 2007, pp. 746–749
3. J. Byrne, Tensilica DSP targets LTE advanced, Mar 2011. http://www.tensilica.com/uploads/pdf/MPR_BBE64.pdf
4. A.P. Chandrakasan, S. Sheng, R.W. Brodersen, Low-power CMOS digital design. IEEE J. Solid State Circuits **27**(4), 473–484 (1992)
5. A. Chattopadhyay, Ingredients of adaptability: a survey of reconfigurable processors. in *VLSI Design*, Jan 2013
6. K. Compton, S. Hauck, Reconfigurable computing: a survey of systems and software. ACM Comput. Surv. **34**, 171–210 (2002)
7. S.H. Fuller, L.I. Millett, Computing performance: game over or next level? Computer **44**(1), 31–38 (2011)
8. M. Garland, et al., Parallel computing experiences with CUDA. IEEE Micro **28**(4), 13–27 (2008)
9. R. Hartenstein, A decade of reconfigurable computing: a visionary retrospective, in *Design, Automation Test in Europe Conference Exhibition (DATE)*, 2001, pp. 642–649
10. J.L. Hennessy, D.A. Patterson, *Computer Architecture: A Quantitative Approach*, 4th edn. (Morgan Kaufmann Publishers, San Francisco, CA, 2003)
11. R.W. Keyes, The impact of Moore's law. IEEE Solid State Circuits Soc. Newslett. **11**(5), 25–27 (2006)
12. T. Lenart, Design of reconfigurable hardware architectures for real-time applications. Ph.D. thesis, Department of Electrical and Information Technology, Lund University, May 2008
13. D. Liu, *Embedded DSP Processor Design: Application Specific Instruction Set Processors*, 1st edn. (Morgan Kaufmann Publishers, San Francisco, CA, 2008)
14. D. Marr, et al., Hyper-threading technology architecture and microarchitecture: a hypertext history. Intel Technol. J. **6**(1), 4–15 (2002)
15. G.E. Moore, Cramming more components onto integrated circuits. Electronics **38**(8), 114–117 (1965)
16. NVIDIA, Tesla C2050/C2070 GPU Computing Processor, July 2010

17. NVIDIA, MATLAB Acceleration on NVIDIA Tesla and Quadro GPUs, 2014
18. K.K. Parhi, *VLSI Digital Signal Processing Systems: Design and Implementation*, 1st edn. (Wiley, New York, 1999)
19. J. Rabaey, *Low Power Design Essentials*. 1st edn. (Springer, New York, 2009)
20. M.B. Taylor, A landscape of the new dark silicon design regime. IEEE Micro **33**(5), 8–19 (2013)
21. R. Tessier, W. Burleson, Reconfigurable computing for digital signal processing: a survey. J. VLSI Signal Process. Syst. **28**, 7–27 (2001)
22. T.J. Todman, G.A. Constantinides, S.J.E. Wilton, O. Mencer, W. Luk, P.Y.K. Cheung, Reconfigurable computing: architectures and design methods. Comput. Digit. Tech. **152**, 193–207 (2005)

Chapter 3
Digital Baseband Processing

Wireless communication has been experiencing explosive growth since its invention. The wireless landscape has been broadened by incorporating more than basic voice services and low data rate transmissions. Taking cellular systems as an example, the fourth generation (4G) mobile communication technology promises to provide broadband Internet access in mobile terminals with up to gigabit-per-second downlink data rate [6]. Compared to the 9.6 kbit/s data services in its 2G predecessor Global system for mobile communications (GSM), 4G systems enhance the data rate by 5 orders of magnitude. This data rate boost is a result of innovations in wireless technology, such as Orthogonal frequency division multiplexing (OFDM) and Multiple-input multiple-output (MIMO). The high speed data links together with advancements in mobile terminals (e.g., phones, tablet computers, and wearable devices) have opened up a whole new world for wireless communication and changed everyone's life. Besides conventional usage like Internet streaming and multimedia playback, interdisciplinary applications like mobile health (mHealth) [19] are emerging. New applications set new demands on wireless services, pushing forward technology developments.

This chapter aims to give a brief description of some modern wireless communication technologies and standards, introduce basic concepts and terminologies used in the rest of the book, and provide an overview of the digital baseband processing tasks in modern systems. Moreover, computational properties of baseband processing tasks are extracted in order to guide hardware developments. Basics of wireless communication, such as symbol modulation and propagation channels, are not addressed but can be found in [13, 16], since the purpose of the present chapter is to highlight design challenges and point out baseband processing properties that can be exploited to achieve efficient hardware implementations. Worth mentioning is that this book mainly focuses on MIMO-OFDM systems because of their importance and popularity in contemporary wireless communication systems. However, support of other wireless technologies is a natural extension and can be easily mapped onto the presented reconfigurable cell array thanks to its flexible hardware infrastructure.

© Springer International Publishing Switzerland 2016   17
C. Zhang et al., *Heterogeneous Reconfigurable Processors for Real-Time
Baseband Processing*, DOI 10.1007/978-3-319-24004-6_3

3.1 Wireless Communication Technologies

To increase the data rate of a wireless system, a straightforward method is to allocate larger bandwidth for data communication. A wide frequency band allows for more data to be transferred at any time. This has been used as one of the main techniques in the transition from 2G to 3G systems, achieving $\sim$40 times data rate speed-up by increasing bandwidth per carrier from 200 kHz to 5 MHz. This trend continues in 4G systems, which further expands bandwidth to 100 MHz with carrier aggregation. However, larger bandwidth increases implementation complexity. This is because multi-path propagation channels are by nature frequency selective [13], and thus affect signals at different frequency bands differently. OFDM technology [5] has been proposed to circumvent the issue of frequency selectivity. To further increase data rate without the expansion of bandwidth, since bandwidth is a limited resource, spatial resources are utilized in addition to time and frequency. MIMO [15] is one such technology that provides various ways of utilizing spatial resources. In modern systems, the two aforementioned technologies are often used together, referred to as MIMO-OFDM systems.

3.1.1 Orthogonal Frequency Division Multiplexing

The key idea of OFDM is to divide a wideband channel into a number of narrowband sub-channels, over which the wideband signal is multiplexed. This way, the frequency response over each of these narrowband sub-channels is flattened, thus reducing the complexity of channel equalization. To enable parallel transmission over flat-fading sub-channels without interfering one another, adjacent narrowband subcarriers need to be separated in frequency (Δf) and arranged such that they are orthogonal to each other. Figure 3.1 illustrates such arrangement. Because of the frequency overlapping, OFDM achieves high spectral efficiency.

In addition to the frequency selectivity, OFDM systems need to cope with channel effects as other wireless systems do. Wireless channels are characterized by multi-path propagation [13]. Signals travelling from one end to the other are reflected, diffracted, and scattered by obstacles, forming multi-path components. Depending on the travelled paths, multi-path components may arrive at the receiver at different time instances. The multi-path propagation will

Fig. 3.1 Orthogonal subcarriers in OFDM

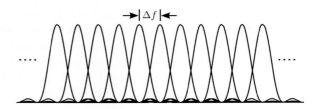

incur interferences between adjacent OFDM symbols, generally referred to as Inter-symbol-interference (ISI). In addition, OFDM systems may suffer from Inter-carrier-interference (ICI), since the orthogonality of the subcarriers may be destroyed by multi-path propagation and imperfections in practical implementations such as carrier-oscillator mismatch.

To avoid both the ISI and ICI, each OFDM symbol is extended with a guard time interval designed to allow channel's impulse response to settle. This guard time interval is filled with a Cyclic prefix (CP), which is a copy of the last part of each OFDM symbol. By discarding CP at the receiver after each symbol reception, given that the CP is long enough to cover the impulse response of the channel, the ISI and ICI can be completely avoided.

3.1.2 Multiple-Input Multiple-Output

MIMO is another important technology in modern wireless communication systems. Compared to single antenna setup, MIMO exploits resources in the spatial domain and provides significant improvements in system capacity and link reliability without increasing bandwidth. In MIMO, three main operation modes exist, namely spatial multiplexing [22], spatial diversity [17], and space division multiple access (also known as multi-user MIMO) [2]. These modes, illustrated in Fig. 3.2, are designed to increase average user spectral efficiency, transmission reliability, and cell spectral efficiency, respectively [11]. To suffice ever-increasing user demands in Quality of service (QoS) while living with the limited bandwidth resources, current trend in wireless systems is to adopt large MIMO dimensions. As an example, the maximum MIMO configuration in the transition from 3GPP Long term evolution (LTE) to its successor LTE-Advanced (LTE-A) is increasing from 4×4 to 8×8, while keeping the bandwidth unchanged.

The benefits of MIMO entail a significant increase in signal processing complexity and power consumption at the receiver, where sophisticated signal processing is required, especially in a fading and noisy channel. For example, Channel state

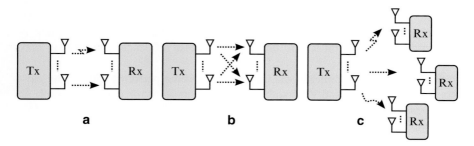

Fig. 3.2 Three operation modes in MIMO, (**a**) spatial multiplexing, (**b**) spatial diversity, (**c**) space division multiple access (multi-user MIMO)

information (CSI) between each pair of transmit and receive antennas should be properly estimated and symbol detection is needed to cancel inter-antenna interferences. As a result, efficient hardware implementation of MIMO receivers has become a critical challenge. Moreover, when combining MIMO with OFDM, it is required to perform the corresponding processing at every OFDM subcarrier, posing even more stringent computational and energy requirements.

3.2   Overview of Digital Baseband Processing

This section introduces baseband processing tasks in MIMO-OFDM systems. Figure 3.3 shows a simplified diagram of a typical MIMO-OFDM transceiver. Note that only digital baseband processing blocks are shown in the figure, whereas the Radio Frequency (RF) front-end and Digital-to-Analog/Analog-to-Digital Converters (DACs/ADCs) are left out.

The receiver (Rx) chain is essentially the reverse processing of tasks performed at the transmitter (Tx). However, the receiver is usually more complex than the transmitter, since it has to reconstruct original data, which may be incomplete and/or distorted during wireless transmission. Some examples of distortions are noise, multi-path channel fading, and imperfections in the RF front-end.

Shaded blocks in Fig. 3.3 are selected in this book as use cases for driving the development of the domain-specific reconfigurable cell array. These blocks are unique to the receiver chain and are key in determining the performance of the entire MIMO-OFDM system.

3.2.1   Channel Encoding/Decoding

The channel encoding block at the transmitter has two main tasks. First, binary data are encoded with error correcting code, such as convolutional codes, which adds redundant information to help receiver detect and correct a limited number of errors without retransmission. Second, encoded data are interleaved to make sure that adjacent bits are not transmitted consecutively in frequency. Interleaving improves transmission robustness with respect to burst errors. Additionally, scrambling is often used to turn the bit stream into a pseudo-noise sequence without long runs of zeros and ones [18].

Opposite to the encoding block, the channel decoder performs data deinterleaving, error correction, and descrambling. Among these, error correction, such as Low-Density Parity-Check (LDPC) code [8], Viterbi [20], and turbo decoding [3], are compute-intensive.

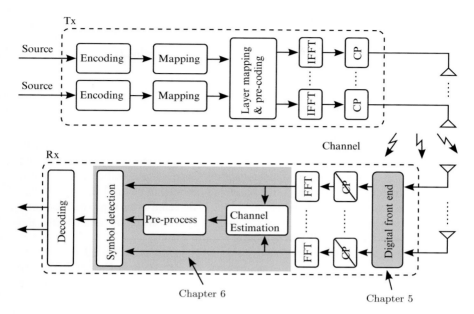

Fig. 3.3 Block diagram of the MIMO-OFDM transceiver. This book focuses on mapping *shaded blocks* onto the dynamically reconfigurable cell array

3.2.2 Symbol Mapping/Demapping

The encoded bit stream is sent for symbol mapping blocks at the transmitter, which are responsible for two tasks. First, in the symbol mapper, the bit stream is mapped to a stream of symbols based on the adopted modulation scheme such as Quadrature amplitude modulation (QAM). Meanwhile, pilots are often added to the symbol stream. Pilots carry known information to both the transmitter and receiver and are used to perform, for example, synchronization and channel estimation at the receiver. Second, the layer mapping block maps the symbol stream onto multiple antennas.

The demapping block (not shown in Fig. 3.3) demaps the symbol stream from multiple antennas, removes pilots, and demaps data-carrying symbols back to the binary bit stream.

3.2.3 Domain Transformation

Before sending data to the analog front-end at the transmitter, symbols from all narrowband subcarriers are collected and are simultaneously transformed to a time-domain signal using an Inverse fast Fourier transform (IFFT). Thereafter, CP is added to each OFDM symbol to protect data transmission from being interfered by ISI and ICI.

At the receiver, CP is removed from each OFDM symbol. Fast Fourier transform (FFT) is used to separate received time-domain signal back to their respective subcarriers.

3.2.4 Digital Front-End Processing

The Digital front-end (DFE) is the first digital processing block in the receiver chain and is responsible for two main tasks [10]. The first is to detect an incoming signal by monitoring the amplitude of signal reception. Once a signal is detected, the DFE wakes up the remaining blocks in the baseband processing chain. Likewise, it puts other blocks into sleep mode when no signal is detected after a pre-defined time interval. The second task is to perform symbol synchronization to determine the exact timing of incoming OFDM symbols.

In addition to the aforementioned tasks, DFE is sometimes used to estimate and/or compensate some of radio impairments [7, 10], such as Carrier frequency offset (CFO), Signal-to-noise ratio (SNR), and IQ imbalance. Chapter 5 presents the mapping of DFE onto the reconfigurable cell array. Target processing tasks include OFDM time synchronization and CFO estimation.

3.2.5 Channel Estimation

To be able to recover transmitted data from the distorted signal reception, it is crucial to have the knowledge on how wireless channel "manipulates" (e.g., attenuates and rotates) the signal transmission. In fact, the performance gain of MIMO-OFDM systems heavily depends on the accuracy of CSI. Channel estimation is used to estimate CSI based on either known information such as pilots and preambles or blind estimation algorithms.

Commonly used channel estimation algorithms are Least square (LS), Minimum mean-square error (MMSE) and its derivatives, FFT, and Singular value decomposition (SVD) estimation. Among these, MMSE estimator provides the highest performance in terms of estimation accuracy, and LS has the lowest computational complexity. The work presented in Chap. 6 adopts an MMSE-based channel estimation algorithm, which provides a balance between performance and computational complexity.

3.2.6 Channel Matrix Pre-processing

The estimated channel matrix at each subcarrier needs to be further processed before being sent to the symbol detector. Depending on the adopted symbol detection algorithm, requirements on channel pre-processing may vary. Commonly used pre-processing algorithms include matrix inversion for linear detectors and QR decomposition (QRD) for tree-search based detectors. Both of these algorithms are used in Chap. 6.

3.2.7 Symbol Detection

In MIMO systems, detection is a joint processing of symbols from all spatial streams, since the symbols all contain a bit of the information after transmitting through the wireless channel. Therefore, the larger the MIMO dimension, the higher the computational complexity is involved in symbol detection. The basic task of a detection is to locate the transmitted data in a constellation diagram. However, since received data are contaminated by channel fading and noise, much effort needs to be spent in the detection process, especially for systems operating at high-order modulation and large antenna numbers.

From the performance point of view, Maximum-likelihood (ML) detection is an optimal detector that solves the closest point search problem. However, ML detector is infeasible to implement due to the exhaustive symbol search that is known to be NP-complete. Popular practical MIMO signal detection algorithms can generally be categorized into two classes, linear and tree-search based detectors, which all have certain performance sacrifice. Linear detection algorithms are preferred for real-time implementations owing to their low computational complexity. Additionally, they are characterized by high Data-level parallelism (DLP), since symbol detection at each spatial stream can be efficiently vectorized and performed in parallel. However, linear detection suffers from huge performance degradation compared to the optimal ML detection, especially for high dimensional MIMO systems. Alternatively, tree-search algorithms are getting much attention because of their near-ML performance. A tree-search detection formulates a minimum-search procedure as a N-depth M-ary complex-valued tree-search problem, where N and M are the number of antennas and constellation size, respectively. Practical suboptimal tree-search detectors solve the NP-complete problem of the optimal ML detection by only traversing through a number of branches. Examples of commonly used algorithms are sphere decoder, K-Best, and their derivatives [1, 4, 9, 12]. One fundamental problem with tree-search algorithms is their intrinsic data dependence between adjacent layers, namely that symbol detection at the ith layer is based on the results of $(i+1)$th layer. Therefore, the native vector structure of MIMO systems is destroyed, resulting in low DLP. In Chap. 6, a vector-level closest point search algorithm in conjunction with linear detectors is introduced, which is highly vector-parallelized, like linear detectors, and at the same time, has the performance close to the level of tree-search detectors.

3.3   Baseband Processing Properties

Based on the analysis of aforementioned digital baseband processing tasks, three computational properties can be observed: *vast complex-valued computing, high data-level parallelism*, and *predictable control flow*. These properties should be exploited during the design of domain-specific reconfigurable cell array to ensure its hardware efficiency.

In the digital baseband processing chain shown in Fig. 3.3, all blocks, except channel encoding/decoding, operate on IQ pairs, which are represented in complex-valued data format. Thus, it is essential to design an Instruction set architecture (ISA) that natively supports complex-valued computing, such as data types, data paths, instruction set, and memory access patterns.

A large portion of computations are performed using vectors, thanks to the parallel-structured MIMO streams. Such computations take place in processing blocks like FFT/IFFT, channel estimation, channel matrix pre-processing, and symbol detection. The abundance of vector processing indicates extensive DLP, which can be utilized to improve processing throughput and reduce control overhead. Moreover, in view of the large number of subcarriers in OFDM, multi-subcarrier processing [21] can be carried out. By performing operations simultaneously on multiple subcarriers, multi-subcarrier processing further exploits DLP in addition to the ones obtained on the algorithm-level. This technique is extensively used in work presented in Chap. 6 and is proven to be useful and effective.

Observed from baseband processing tasks, there is no or little backward dependency between one another [14]. This makes control flow predictable and can therefore simplify the control path to reduce overhead.

References

1. L.G., Barbero, J.S. Thompson, Fixing the complexity of the sphere decoder for MIMO detection. IEEE Trans. Wirel. Commun. **7**(6), 2131–2142 (2008)
2. G. Bauch, G. Dietl, Multi-user MIMO for achieving IMT-advanced requirements, in *International Conference on Telecommunications (ICT)*, June 2008, pp. 1–7
3. C. Berrou, A. Glavieux, Near optimum error correcting coding and decoding: turbo-codes. IEEE Trans. Comput. **44**(10), 1261–1271 (1996)
4. A. Burg, et al., VLSI implementation of MIMO detection using the sphere decoding algorithm. IEEE J. Solid State Circuits **40**(7), 1566–1577 (2005)
5. R.W. Chang, Synthesis of band-limited orthogonal signals for multichannel data transmission. Bell Syst. Tech. J. **45**(10), 1775–1796 (1966)
6. E. Dahlman, S. Parkvall, J. Skold, *4G: LTE/LTE-Advanced for Mobile Broadband*, 1st edn. (Academic, New York, 2011)
7. I. Diaz, Algorithm-architecture co-design for digital front-ends in mobile receivers. Ph.D. thesis, Department of Electrical and Information Technology, Lund University, 2014
8. R.G. Gallager, Low-density parity-check codes. IRE Trans. Inf. Theory **8**(1), 21–28 (1962)
9. Z. Guo, P. Nilsson, Algorithm and implementation of the K-best sphere decoding for MIMO detection. IEEE J. Sel. Areas Commun. **24**(3), 491–503 (2006)

10. F. Horlin, A. Bourdoux. *Digital Compensation for Analog Front-Ends: A New Approach to Wireless Transceiver Design*, 1st edn. (Wiley, 2008)
11. L. Liu, J. Löfgren, P. Nilsson, Area-efficient configurable high-throughput signal detector supporting multiple MIMO modes. IEEE Trans. Circuits Syst. Regul. Pap. **59**(9), 2085–2096 (2012)
12. M. Li, et al., Optimizing near-ML MIMO detector for SDR baseband on parallel programmable architectures. in *Design, Automation and Test in Europe (DATE)*, Mar 2008, pp. 444–449
13. A.F. Molisch, *Wireless Communications*, 2nd edn. (Wiley, New York, 2010)
14. A. Nilsson. Design of programmable multi-standard baseband processors. Ph.D. thesis, Department of Electrical Engineering, Linköping University, 2007
15. A. Paulraj, R. Nabar, D. Gore, *Introduction to Space-Time Wireless Communications*, 1st edn. (Cambridge University Press, Cambridge, 2008)
16. J. Proakis, M. Salehi, *Digital Communications*, 5th edn. (McGraw-Hill Science, New York, 2007)
17. C. Spiegel, J. Berkmann, Z. Bai, T. Scholand, C. Drewes, MIMO schemes in UTRA LTE, a Comparison, in *IEEE Vehicular Technology Conference (VTC)*, May 2008, pp. 2228–2232
18. E. Tell, Design of programmable baseband processors. Ph.D. thesis, Department of Electrical Engineering, Linköping University, 2005
19. United Nations Foundation and The Vodafone Foundation, mHealth for development: the opportunity of mobile technology for healthcare in the developing world, 2011
20. A.J. Viterbi, Error bounds for convolutional codes and an asymptotically optimum decoding algorithm. IEEE Trans. Inf. Theory **13**(2), 260–269 (1967)
21. C. Yang, D. Marković, A flexible DSP architecture for MIMO sphere decoding. IEEE Trans. Circuits Syst. Regul. Pap. **56**(10), 2301–2314 (2009)
22. C. Yuen, B.M. Hochwald, Achieving near-capacity at low SNR on a multiple-antenna multiple-user channel. IEEE Trans. Commun. **57**(1), 69–74 (2009)

Chapter 4
The Reconfigurable Cell Array

Emerging as a prominent technology, reconfigurable architectures have the potential of combining high hardware flexibility with high performance data processing. Conventional fine-grained architectures, such as Field-programmable gate arrays (FPGAs), provide great flexibility by allowing bit-level manipulations in system designs. However, the fine-grained configurability results in long configuration time and poor area and power efficiency, and thus restricts the usage of such architectures in time-critical and area/power-limited applications. To address these issues, recent work focuses on coarse-grained architectures, aiming to provide a balance between flexibility and hardware efficiency by adopting word-level data processing. In this chapter, a coarse-grained dynamically reconfigurable cell array architecture is introduced. The architecture is constructed from an array of heterogeneous functional units communicating via hierarchical network interconnects. The strength of the architecture lies in simplified data sharing achieved by decoupled processing and memory cells, substantial communication cost reduction obtained by a hierarchical network structure, and fast context switching enabled by a unique run-time reconfiguration mechanism. The presented reconfigurable cell array serves as a baseline architecture for two case studies presented in Chaps. 5 and 6.

4.1 Introduction

The evolution of user applications and increasingly sophisticated algorithms call for everincreasing performance of data processing. Meanwhile, to prolong system's operating time of battery operated devices, contemporary designs require low power consumption. A typical example is baseband processing in 4G mobile communication, which demands a computational performance of up to 100 Giga operations per second (GOPS) with a power budget of around 500 mW in a single user terminal [11]. In addition to computational capability and power consumption,

© Springer International Publishing Switzerland 2016
C. Zhang et al., *Heterogeneous Reconfigurable Processors for Real-Time Baseband Processing*, DOI 10.1007/978-3-319-24004-6_4

flexibility becomes an important design factor, since system platforms need to cope with various standards and support multiple tasks simultaneously. Therefore, it is no longer viable to dedicate a traditional application-specific hardware accelerator to each desired operation, as the accelerators are rather inflexible and costly in system development, validation, and maintenance (e.g., bug-fixes and function updates).

To achieve a balance among the aforementioned design requirements, reconfigurable architectures have gained increasing attention from both industry and academia. These architectures enable hardware reuse among multiple designs and are able to dynamically allocate a set of processing, memory, and routing resources to accomplish current computational demands. Moreover, reconfigurable architectures allow mapping of future functionality without additional hardware or manufacturing costs. Therefore, by using platforms containing reconfigurable architectures it is possible to achieve high hardware flexibility while sufficing the stringent performance and power demands [35].

Fine-grained and coarse-grained arrays are two main variants of reconfigurable architectures. While the former has the ability to map any logic functions at bit-level onto their fine-grained lattice, the latter is constructed from larger building blocks in a size ranging from Arithmetic logic units (ALUs) to full-scale processors. Compared to fine-grained architectures, the increased granularity in Coarse-grained reconfigurable architectures (CGRAs) reduces routing area overhead, improves configuration time, and achieves higher power efficiency despite less mapping flexibility.

This chapter introduces a coarse-grained dynamically reconfigurable cell array architecture, which will be used as a design template in the remaining chapters of the book. The cell array is a heterogeneous CGRA, containing an array of separated processing and memory cells, both of which are global resources distributed throughout the entire network. Array elements communicate with one another via a combination of local interconnects and a hierarchical routing network. All array elements are parameterizable at system design-time, and are dynamically reconfigurable to support run-time application mapping. The following summarizes distinguished features of the architecture.

- The heterogeneity of the architecture allows integration of various types of resource cells into the array.
- Separation of processing and memory cells simplifies data sharing among resource cells.
- A hierarchical Network-on-chip (NoC) structure combines high-bandwidth local communication with flexible global data routing.
- In-cell resource reconfigurations conducted by distributed processing cells enable fast run-time context switching.

It should be pointed out that the presented cell array is a general architecture, which can be, in principle, used to map any algorithms, tasks, and applications. However, this book mainly focuses on signal processing in wireless communication, more specifically, digital baseband processing at the receiver. By exploiting computational properties of the target application domain, various architectural improvements can be carried out on the baseline architecture to further improve hardware performance and efficiency. Improvements will be illustrated through case studies in Chaps. 5 and 6. The present chapter serves to give an overview of the cell array, including the overall architecture, basic functionality and framework of each resource cell, network infrastructure, hardware reconfigurability, and design methodology.

The remainder of this chapter is organized as follows. Section 4.2 discusses related work with a focus on architectures designed specifically for digital wireless communication. Section 4.3 introduces the reconfigurable cell array architecture, presents details of each resource cell, and describes different ways of managing system configurations. Section 4.4 presents design flow for constructing a reconfigurable cell array. Section 4.5 summarizes this chapter.

4.2   Prior Work and State-of-the-Art

A number of reconfigurable architectures have been proposed in open literature for a variety of application domains [1, 10, 12, 18, 30, 35]. Presented architectures are characterized with various design parameters, such as granularity, processing and memory organization, coupling with a host processor, communication fabric, reconfigurability, and programming methodology. Describing each of the architectures with respect to those parameters is cumbersome and is in fact unnecessary because of architectural similarities. Instead, previously proposed architectures are classified into three broad categories based on the coupling between processing and memory units and their interconnects. In addition, the following discussions are restricted to architectures designed specifically for digital baseband processing in wireless communication. These systems are sometimes referred to as baseband processors [25] or Software-defined radio (SDR) platforms [14]. The three classes of reconfigurable architectures are illustrated in Fig. 4.1.

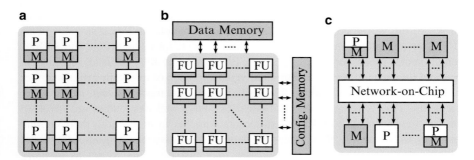

Fig. 4.1   Three classes of reconfigurable architectures, (**a**) homogeneous processor array, (**b**) Function unit (FU) cluster, (**c**) heterogeneous resource array

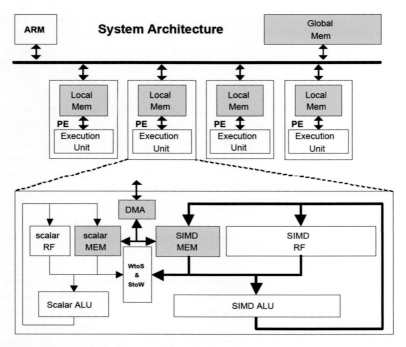

Fig. 4.2   Overview of SODA architecture [23]

The first group of architectures (Fig. 4.1a), such as the PicoArray [4] from Picochip, the Signal processing On Demand Architecture (SODA) platform [21, 23] from the University of Michigan, Ann Arbor, Cadence's ConnX BaseBand Engine (BBE) [9], and Ninesilica platform [2] are constructed from an array of homogeneous processors. Each processor has exclusive access to its own memory. As an example, Fig. 4.2 shows an instance of SODA architecture. It is made up of four cores, each containing asymmetric dual pipelines, for scalar and Single instruction multiple data (SIMD) execution, and scratchpad memories. According to [1, 6], homogeneous architectures are not cost effective in supporting algorithms in which the workload cannot be balanced among multiple Processing Elements (PEs), or algorithms involving hybrid data computations like scalar and various length vector processing. In these cases, PEs cannot be fully utilized, resulting in reduced hardware efficiency. Additionally, the approach of integrating data memory inside PE results in difficulties when sharing data contents between surrounding elements. This is because PEs at both data source and destination are involved in data transmissions to load and store data contents from and into their internal memory. Consequently, these inter-core data transfers may take a significant amount of processing power, and in some cases may just turn PEs to act as memory access controllers, reducing hardware usage. Moreover, storage capacities of data memories inside PEs are fixed after chip fabrication, which may reduce the flexibility and applicability of platforms.

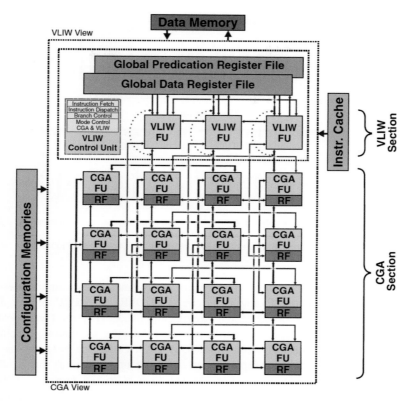

Fig. 4.3 ADRES instance with 16 Coarse-Grain Array (CGA) Function units (FUs) and three Very long instruction word (VLIW) FUs [17]

Architectures in Fig. 4.1b are built from atomic Function units (FUs), named as FU cluster. Examples of this group are the Architecture for Dynamically Reconfigurable Embedded Systems (ADRES) [8] from IMEC, NXP's EVP16 processor [36], and the eXtreme Processing Platform (XPP) [5] from PACT Informationstechnologie. Figure 4.3 illustrates an ADRES instance, which contains a Coarse-Grain Array (CGA) of FUs and three Very long instruction word (VLIW) FUs. The VLIW FUs and a limited subset of the CGA FUs are connected to globally shared data Register Files (RFs), which are used to exchange data between the two sections. For this group of architectures, memory accesses may suffer from long-path data transfers, since data memories are accessible only from the border of the cluster. These long-path transfers may result in high data communication overhead especially for large-size clusters. Additionally, centralized memory organization may cause memory contention during concurrent data accesses, which may become a bottleneck for high dimensional computations (e.g., vector processing).

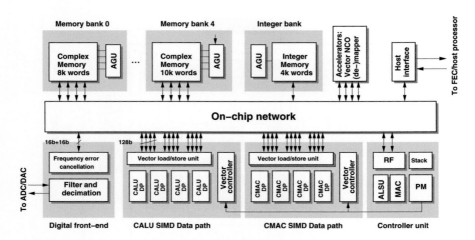

Fig. 4.4 Overview of Coresonic Single instruction stream multiple tasks (SIMT) DSP [24]

Figure 4.1c shows architectures consisting of heterogeneous units interconnected through an on-chip network. Examples are the SIMT DSP [24] from Coresonic, the FlexCore [34], the Transport Triggered Architecture (TTA) [19], the Dynamically Reconfigurable Resource Array (DRRA) [29], and Adaptive Computing Machine (ACM) [27] from Quicksilver. The SIMT DSP from Coresonic is shown as an example in Fig. 4.4, where multiple vector execution units, memory banks, and application-specific accelerators are connected to a restricted crossbar switch. Because of the heterogeneity, this type of architectures can be tailored to specific application domains to achieve efficient computations. However, one potential problem is the overhead of network interconnects, which increases linearly with the number of array nodes. This may restrict the usage of architectures in high dimensional data applications. For instance, the customized network presented in a 16-bit architecture [19] consumes almost the same area as all its arithmetic parts. Thereby, extensions to vector processing using many more array nodes may be unaffordable. Additionally, when considering hybrid computing, various-width data transfers via shared homogeneous network interconnects are not cost effective and may require frequent data alignment operations. Moreover, architectural scaling may require redesign of network interconnects, resulting in poor scalability.

In view of the high hardware efficiency owing to heterogeneity, the cell array is built upon the third architectural category, the heterogeneous resource array. To tackle the NoC overhead and scalability issue, a hierarchical network topology is adopted, which contains high-bandwidth local interconnects and flexible global data routing. Additionally, to ease data sharing between surrounding array elements, processing and memory cells are separated as two distinct function units which are shared as global resources and distributed throughout the entire network. The following sections present the cell array architecture in detail.

4.3 Architecture Overview

The reconfigurable cell array is constructed from heterogeneous tiles, containing any size, type, and combination of resource cells. As an example, a 4-tile cell array is shown in Fig. 4.5. Besides the cell array, the entire system platform contains a master processor, a Multi-port memory controller (MPMC), a Stream data controller (SDC), a Stream configuration controller (SCC), and a number of peripherals. The master processor schedules tasks to both the cell array and peripherals at run-time. The MPMC interfaces with external memories, while the SDC and SCC supply the cell array with data and configurations, respectively. Since the focus of this book is on the cell array, other system blocks will not be discussed.

Resource cell (RC) is a common name for all types of functional units inside the cell array, including *processing*, *memory*, and *network routing cells*. Within the cell array, processing and memory cells are separated as two distinct functional units. This arrangement has following advantages.

- **Easy data sharing:** The separation of memory from processing cells signif-icantly simplifies data sharing, as memory cells can be shared by multiple processors without physically transferring data. Memory coherence is preserved by allowing direct data transfers between memory cells without involving processors.
- **Flexible memory usage:** Memory cells can be individually configured to provide different access patterns, such as First in first out (FIFO), stack, and random access.
- **Advanced data access control:** Processing cells can be used as Direct memory access (DMA) controller to accomplish irregular and/or advanced memory access, e.g., bit-reversal in FFT/IFFT.

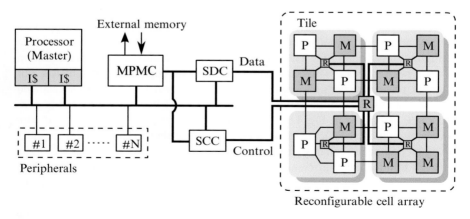

Fig. 4.5 Overview of system platform containing a coarse-grained dynamically reconfigurable cell array. The separated processing (P) and memory cells (M) communicate over a hierarchical network using both local interconnects and global routing with network routing cells (R)

- **Dynamic memory allocation:** When larger memory capacity is required than a single cell can provide, multiple memory cells can be concatenated at system run-time to provide larger data storage.
- **Simplified programming model:** The separated data processing and memory access management naturally supports dataflow programming languages like CAL [16] and computation models like Kahn process networks [20]. Additionally, changes of execution clock cycles in either processing or memory cells have no influence on the control flow of the entire cell array architecture, since inter-cell communication is self-synchronized through NoC data transfers.
- **Natural support for future technology:** The advantage of processing and memory separation may be more pronounced when using 3D stacking technology [26]. With this technology support, processing and memory cells can be placed at different chip layers and interconnected by using Through-Silicon Vias (TSVs). This approach may further increase memory access bandwidth, reduce delays of network interconnects, and ease chip layout and routing process.

To meet the computational and flexibility requirements while keeping a low control overhead, a variety of functional units are integrated. Communication between RCs is managed by combing local interconnections for high data rate and a global routing network for flexibility. Compared to other interconnect topologies, the hierarchical network provides tighter coupling to heterogeneous RCs. For instance, connections within each tile can be localized to suffice both bandwidth and efficiency requirements, while hierarchical links provide flexible routing paths for inter-tile communication. All RCs in the array are configured dynamically on a per-clock-cycle basis, in order to efficiently support run-time application mapping.

4.3.1   Processing Cell

Processing cells contain computational units to implement algorithms mapped on the cell array. Additionally, they can be used to control operations and manage configurations of other RCs. The heterogeneity of the architecture allows integration of any type of processors in the array to suffice various computational demands. For example, processing cells may be built as general-purpose processors or specialized functional units. Each processing cell is composed of two parts, processing *core* and *shell*, illustrated in Fig. 4.6a. Encapsulated by the processing shell, the core interfaces with other RCs via network adapters in the shell. Thanks to this modular structure, processing cell customization is simplified since only the core needs to be replaced to implement different computational operations. In addition, integration of customized functional units, either user-defined Register transfer level (RTL) or licensed Intellectual property (IP) cores, is supported in the cell array without changing the network interface. The network adapters in the processing shell are mapped as registers that are directly addressable by the core. The number of adapters

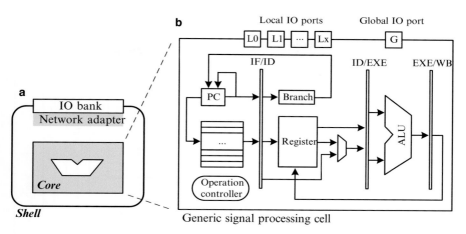

Fig. 4.6 (**a**) Block diagram of a processing cell, consisting of a *core* and a *shell*. (**b**) Architecture of a Generic signal processing cell (GPC)

in a processing cell is parameterizable at system design-time. For illustration, the following presents a generic signal processing cell, named as GPC for short, which has been used in a flexible Fast Fourier transform (FFT) core presented in [38].

A GPC is a customized Reduced instruction set computing (RISC) processor with enhanced functionality for digital signal processing and support for fast network port access. An overview of processor pipeline stages and internal building blocks is shown in Fig. 4.6b. Communication I/O ports are mapped as registers, directly accessible in the same way as General purpose registers (GPRs). An instruction that accesses I/O port registers is automatically stalled until data become available. Hence, additional operations to move data between GPRs and I/O ports are avoided. The GPC performs data memory operations by connecting with one or more memory cells via communication I/O ports. Using direct I/O addressing, load-store and computational operations may be combined into one instruction. Consequently, clock cycles associated with memory operations are eliminated in contrast to conventional load-store architectures. Moreover, the implicit load-store operations lead to a compact code size, and make the memory operations possible in all instructions. Enhanced functionalities for digital signal processing include multiply-accumulate, radix-2 butterfly, and data swap. To reduce control overhead in computationally intensive inner loops, the GPC includes a zero-delay Inner loop controller (ILC). The ILC comprises a special set of registers that are used to store program loop count and return address. During program execution, the loop operation is indicated by an end-of-loop flag annotated in the last loop instruction. The operation mode and status of each processing cell can be controlled and traced conditionally during run-time. For example, it is possible to halt instruction execution, step through a program segment, and partially load a program. Because

Fig. 4.7 Architectural block diagram of a memory cell

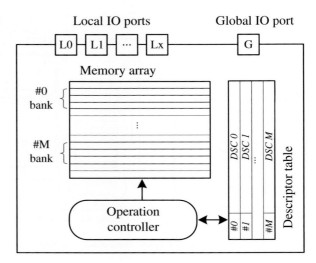

of the simple pipeline structure and enhanced data processing operations, the GPC may be used for regular data processing and control-flow handling, such as linear filter, FFT/IFFT, and DMA control.

4.3.2   Memory Cell

The distributed memory cells provide both processing cells and data communication with shared storage to allow buffering, reordering, and synchronization of data. Each memory cell contains a *memory array*, a *DeSCriptor (DSC) table*, and an *operation controller*, as illustrated in Fig. 4.7. The memory array can be dynamically configured to emulate one or more memory banks, while the DSC table is an array of configuration registers containing user-defined memory operations. All stored DSCs are dynamically configurable, and may be traced back for debugging. Each DSC is 64-bit long, which defines the size and operation mode of a memory bank, records memory operation status, and specifies I/O ports for stream transfers. The configuration options of a DSC are listed in Table 4.1. The 64-bit DSC is composed of two 32-bit parts that are individually configurable. Specifying memory operations using DSCs relieves processing cells from memory access managements, resulting in reduced control overhead and improved processing efficiency. The operation controller manages and schedules DSC execution, monitors data transactions, and controls the corresponding memory operations.

Using memory DSCs, each memory bank can be configured to emulate either FIFO or Random access memory (RAM) behavior. In the FIFO mode, the allocated memory bank operates as a circular buffer. Address pointers are managed by the operation controller and are automatically increased each time the DSC is executed.

Table 4.1 Example of a memory DeSCriptor (DSC)

	Field	Bits	Description
Part I	dtype	31-30	Operation mode select
	rd_ok/active	29	FIFO reading status/RAM active transfer flag
	wr_ok/rnw	28	FIFO writing status/RAM read–write select
	src/paddr	27-24	FIFO data source port/RAM address port
	dst/pdata	23-20	FIFO data destination port/RAM data port
	id	19-10	Global packet destination ID
	Base	9-0	Start address
Part II	High	31-22	End address
	rptr/ptr	21-12	Current FIFO reading pointer/RAM data pointer
	wptr/tsize	11-2	Current FIFO writing pointer/RAM data transfer size
	io_bank_rst	1	I/O port register reset
	Reserved	0	Reserved

Multiple memory cells operating in FIFO mode may be cascaded to form one large data array. This feature provides flexible memory usage, and reduces unit capacity requirement as well as hardware footprint in a single memory cell. Additionally, since large storage may lead to an irregular physical memory layout, slicing it into smaller modules eases hardware placement and routing. When operating memory cell is in the RAM mode, a data service request (read or write) is required to specify the start address and data transfer size. The operation controller is responsible for keeping track of data transfers, managing memory address pointers, and updating DSCs. Conditions to execute a memory DSC are resolved by inspecting both incoming and outgoing packet transfers and current memory status. For example, writing data to a full FIFO will not be executed until at least one data is read.

The length of the DSC table, the size of the memory array, and the number of local I/O ports are configurable at system design-time, while memory descriptors are dynamically reconfigurable.

4.3.3 Network-on-Chip

To enable communication between any pair of resource cells, most existing NoCs are based on flexible interconnect topologies, such as 2D-mesh, spidergon, and their derivatives [13, 28, 33, 37]. Additionally, various routing algorithms (e.g., static and dynamic) and switching techniques (e.g., wormhole and Time Division Multiplexing (TDM)) are employed to reduce traffic congestions, provide service guarantees, and shorten communication latency [7]. Although most NoC implementations can suffice performance requirements with respect to latency and bandwidth, they often appear to be area and power consuming. For instance, NoCs used in [15, 19] take almost the same area as all their logic parts and the one in [15] consumes about 25 % of total power.

Fig. 4.8 An overview of the hierarchical NoC arranged in a 4 × 4 array. Neighboring RCs are directly communicated via local interconnects, while global transfers, *shaded in grey*, are achieved through hierarchical routing with tree-structured network router (R)

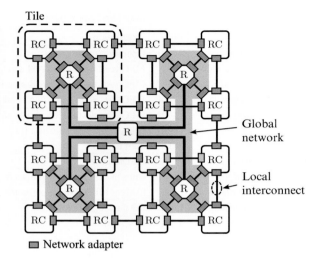

This book aims at developing an area and power efficient NoC by fully exploiting the property of *communication locality* in reconfigurable architectures [37]: data traffic is mostly among nearest neighbors (referred to as local communication), while long distance (global) transfers that require routing supervisions are of a small portion. Therefore, the primary concern of local network design should be on high bandwidth and low cost, while simple routing and switching techniques are sufficient for global transfers to provide adequate flexibility support.

In light of the aforementioned property, a hierarchical network architecture is adopted that splits local and global communication into two separate networks, which are handled independently using different network topology and switching techniques. Figure 4.8 illustrates an overview of the hierarchical NoC deployed in a 4 × 4 array. Communication between neighboring RCs (local) within each tile is performed using bi-directional dedicated links, whereas inter-tile global transfers are realized through a hierarchy of network routers structured in a tree topology using a static routing strategy, see shaded part in Fig. 4.8. Thanks to the network separation and hierarchical arrangement, the presented NoC can be easily scaled by extending tree hierarchies of the global network and neighboring local interconnects without affecting others. Additionally, in conjunction with the tile-based architecture, the adopted NoC intrinsically supports Globally asynchronous locally synchronous (GALS) network construction. For example, synchronous transfers are performed within each tile and the global network (together with additional asynchronous FIFOs) is used to bridge between different clock domains.

To connect RCs to the local and global network, adapters are used as a bridge between high level communication interfaces employed by RCs and network specific interfaces implemented in the NoC. In this book, AMBA 4 AXI4-stream protocol [3] is adopted for implementing NoC adapters.

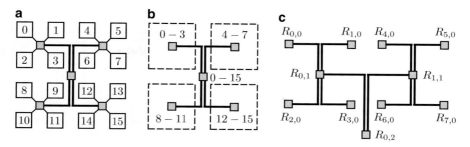

Fig. 4.9 A simplified view of the global network in a 4×4 array. (**a**) Each RC is labeled with a unique network ID. (**b**) A range of consecutive IDs (base-high) are assigned to each static routing table. (**c**) Hierarchical router naming as $R_{\text{index,level}}$

Global Network

Global network enables non-neighboring nodes to communicate within the array and provides an interface to external blocks, e.g., memory and master processor. Figure 4.9a depicts a simplified view of the global network in a 4×4 cell array, wherein each RC is labeled with a unique network IDentifier (ID) used as a routing address for global data transfers. Global communication is based on packet switching and carried out using a hierarchy of tree-structured network routers. Routers forward data packets based on a static routing lookup table that is generated at design-time in accordance to physical network connections. For example, the router in the upper-left tile shown in Fig. 4.9b forwards packets to RCs with IDs ranging from 0 to 3. In the tree-structured global network, each router is denoted as $R_{i,l}$, where i is the router index number and l is the router hierarchical level, illustrated in Fig. 4.9c. A link from a router $R_{i,l}$ to $R_{i,l+1}$ is referred to as an uplink. Any packet received by a router is forwarded to the uplink router if the packet destination ID falls outside the range of the routing table. Since routing network is static, there is only one valid path from each source to each destination. This simplifies network traffic scheduling, reduces hardware complexity, and enables each router instance to be optimized individually during hardware synthesis. However, a drawback is network congestion compared to adaptive routing algorithms. Systematic analysis on network performance at design-time is therefore crucial to avoid traffic overload. However, considering the high communication locality, this congestion issue is of less concern and does not hinder the performance of the presented NoC. For network traffic modeling and performance evaluation, a SystemC-based exploration environment SCENIC can be used. Details of SCENIC can be found in [22, 31, 32].

Fig. 4.10 (**a**) Block diagram
of the network router. (**b**)
Internal building blocks of a
decision unit

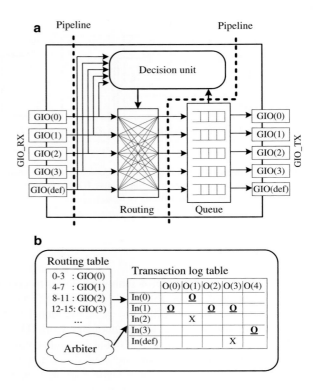

Network Routing Cell

Network router forwards data packets over the global routing network. Each router
consists of three main building blocks: a decision unit, a routing structure, and an
output packet queue, see Fig. 4.10a. In each clock cycle, the decision unit monitors
incoming and outgoing packets, looks up the routing path, handles data transfers,
and configures the routing structure to forward data packets accordingly. The routing
structure is made up of a full-connection switch, capable of handling multiple
data requests in each clock cycle. The output packet queue, operating in a FIFO
basis, buffers data packets travelling through the global network. The depth of the
output queue and FIFO type (either synchronous or asynchronous) are design-time
configurable, used to suffice different NoC requirements.

Figure 4.10b shows an overview of the decision unit inside each network router.
It contains a static routing lookup table, a transaction log table, and a packet arbiter.
Every arriving data packet is checked and recorded in a transaction log table, marked
with 'X' in Fig. 4.10b. The logged transactions are prioritized and handled based on
different arbitration policies and conditions of output queues. Two simple arbitration
policies are currently supported and are design-time configurable: the fixed and
round-robin scheme. With fixed arbitration, the arbiter always starts from the first
log entry, and traverses column-wise through the entire table until a candidate

transaction is found, marked with 'O' in Fig. 4.10b. A transaction is considered to be a candidate when it is recorded in the log table and the corresponding output packet queue is not full. With this approach, all transactions are assigned with priorities according to their position in the log table. In contrast, the round-robin algorithm provides a starvation-free arbitration, which assigns time slices to each entry in equal portions and handles all transactions in order without priority. After routing arbitration, selected candidate transactions (marked with '$\underline{O}$'s in Fig. 4.10b) are forwarded to the corresponding output queue in the following clock cycle. Considering delays caused by input I/O register, pipelined routing operations, and output FIFO, packet forwarding through each network router induces 3 Clock Cycles (CCs) transport latency (without Tx I/O register).

Local Network

Local network consists of dedicated interconnects between neighboring RCs (Fig. 4.8). Thus, local transfers require no routing supervision and provide guaranteed throughput and transport latency. Compared to nearest-neighbor transfers in conventional mesh-based networks, bandwidth overhead due to redundant traffic headers is completely avoided in the adopted local communication. For example, considering the illustrated 4×4 array (Fig. 4.8) and a simple XY routing in mesh networks, it is required to have at least 4 bits in the traffic header to indicate destination coordinates of both X and Y directions. This overhead is more pronounced when sophisticated routing algorithms are used, such as source and adaptive routing. Therefore, avoiding such bandwidth overhead in every neighboring data transfer contributes to total NoC efficiency.

Communication Flow Control

To assure safe delivery and self-synchronization of each data transfer, flow control is used in both local and global networks, implemented using a FIFO-like handshake protocol, illustrated in Fig. 4.11. In addition to conventional valid-ACKnowledgement (ACK) handshaking, FIFO-like operations are adopted to reduce communication latency. The basic idea is to use I/O registers as eager transport buffers, which are writable as long as the buffers are not full. A communication link is suspended only if all buffers are fully used, transmitter has more data to send, and receiver has not yet responded to previous transfers. As indicated in Fig. 4.11, the ACK signal in each I/O register has two acknowledgement mechanisms. In the case of empty buffers, data transfers are automatically acknowledged by the I/O registers. Otherwise, the ACK signal is driven by the succeeding data receiver. This way, the ACK signal acts as an empty flag of the transport buffers and reflects the status of the communication link. Data transmitter can proceed with other operations immediately the ACK from the succeeding stage is received, without waiting for

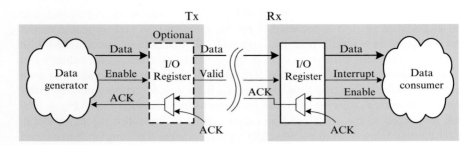

Fig. 4.11 Data communication flow control with FIFO-like handshake protocol. I/O registers at the transmitters are design-time configurable

Table 4.2 Summary of the hierarchical NoC

	Topology	Switching	Latency	Throughput
Local	Direct link	Circuit (GS)	1 CC	1 CC
Global	Tree	Packet (BE)	4 CCs/hierarchy	1 CC/hop

the final destination to respond. Compared to conventional end-to-end handshake protocols, in which the ACK signal is sent all the way from the final destination and requires multiple clock cycles to propagate through all I/O registers, the adopted scheme divides long communication path into smaller segments (hops), each having 1 CC transport latency under the case of eager receiver. As for communication between neighboring RCs, the FIFO-like handshaking results in at least two times latency reduction in comparison to the end-to-end scheme.

To sum up, Table 4.2 lists the characteristics of the presented NoC. Benefiting from the dedicated interconnects, local network provides Guaranteed Services (GS) and has a transport latency and throughput of 1 CC. The global network offers Best-Effort (BE) packet switching and induces additional 3 CCs transport latency (without Tx I/O register) every time a network router is used. However, the throughput of global transfers via routers is still 1 CC thanks to the pipelined architecture.

4.3.4   Resource Configuration

Dynamic reconfigurations for all RCs are managed in two ways, either by a master processor via hierarchical network or by any of the processing cells distributed in the cell array. When configuring RCs through the master processor, see Fig. 4.5, an SCC is used to assist network packet transfers. The SCC contains a stream table programmed by the master processor, and provides information about where and how network packets should be transmitted. For each configuration, the SCC loads data from external memory via the MPMC, packs data as network packets, and transfers the packets to target RC via the hierarchical routing network. The advantages of configuring RCs from a centralized master processor are twofold. First, as configurations are loaded from external memories, requirement for the size of configuration files is reduced. Second, the master processor may utilize hardware resources efficiently on a system level, based on the needs of application mapping. After receiving a task, the master processor assesses the computational workload, checks the status of RCs, and assigns the task to achieve maximum efficiency. For instance, the master may partition and assign the task to different RCs or time-multiplex a single RC. A drawback of the centralized configuration is the communication latency through the global network, as mentioned in Sect. 4.3.3. As a result, the centralized RC configuration scheme is mainly used for transferring large configuration files to the cell array during, for example, context switching between different application mappings.

RC configurations and supervisions can also be conducted inside the cell array by using distributed processing cells. This is achieved by storing RC configurations as special instructions locally inside processing cells, which transmit configuration packets to the corresponding RCs during program execution. With this approach, run-time configurations are smoothly integrated into the normal processing flow. For example, configurations are issued immediately the current task is completed without interrupting and waiting for responses from an external host. Additionally, configuration packets are transmitted mainly using local interconnects, avoiding long configuration latency due to global communication. Because RC configurations are stored as part of the local programs of processing cells, configuration file size needs to be kept down when using this approach. Therefore, this in-cell configuration scheme is suitable for small function configurations in the cell array, such as adapting algorithms for different standards or operating scenarios.

4.4   Design Flow

Constructing a reconfigurable cell array generally involves three design phases (Fig. 4.12), *specification*, *design*, and *implementation*, and three design methodologies, *algorithm–architecture*, *hardware–software*, and *processing–memory co-design*. In the specification phase, target wireless communication standards and baseband processing tasks need to be defined first. This is to limit the scope of the development, in order to enable design-space exploration for creating an efficient architecture. To this end, the target standards and tasks should have some common computational characteristics to enable hardware sharing and acceleration, otherwise diverse operation requirements would lead to a generic architecture with low hardware efficiency. After defining the standards and tasks to cover as well as performance specifications, algorithm selection and operation analysis are carried out to provide a good foundation for the following hardware development. Examples of selection criteria are computational parallelism, precision requirement, and regularity of operations. To make use of essential architectural characteristics,

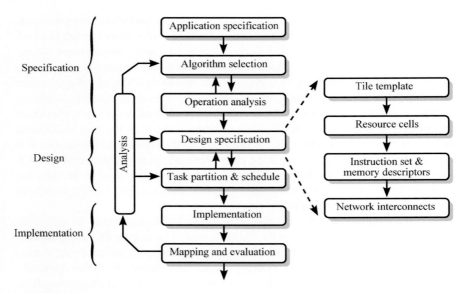

Fig. 4.12 Data flow of constructing a reconfigurable cell array

algorithms often need to be tailored for the target hardware architecture. This is referred to as *algorithm–architecture co-design*, which is usually an iterative process for obtaining an optimum result. It should be pointed out that algorithm selection is especially important when handling multiple standards and tasks, since it has a great influence on efficient usage of underlying hardware, such as resource sharing over time.

The design phase involves two steps, *design specification* and *task scheduling*. After obtaining a rough estimation on computational requirements and memory usage from the operation analysis, design specifications for constructing the cell array are elaborated. These include design of resource tiles, RC selections, instruction set and memory descriptors, and NoC interconnects (see the sub-figure shown on the right side of Fig. 4.12). First, the number of resource tiles is determined based on the computational requirements and design constraints such as area and timing budget. Besides, a rough task scheduling is performed on the tile-level to partition tasks to resource tiles. Second, RC selection is carried out for each tile based on the rough task partition. The number of RCs assigned to each tile is determined again by the computational requirements and design constraints. Third, instruction set and memory descriptors are specified. The number of instructions and memory descriptors is a trade-off between the ease of software implementation and hardware complexity [25]. This is referred to as *hardware–software co-design*. Last, network interconnects may be refined to further improve hardware efficiency. Examples include bandwidth enhancements for heavy-traffic links and pruning of unused network connections. With the elaborated design specifications, detailed task partition and scheduling can be carried out, requiring detailed operation analysis to better assign tasks to RCs, namely processing and memory cells. This is referred to as *processing–memory co-design*. For example, memory address manipulations, such as stride access and matrix transpose, are better performed in memory cells by using descriptor specifications, since no physical memory access (read & write) is needed. Moreover, analysis of network traffic is required to avoid problems like network congestion.

Once the design specifications and the task scheduling are completed, the cell array is implemented and target tasks are mapped. Mapping results are evaluated and fed back to the corresponding design phase for further improvements in case the design requirements are not met.

4.5   Summary

This chapter introduces the coarse-grained dynamically reconfigurable cell array architecture, aiming to provide a balance among performance, hardware efficiency, and flexibility. The presented architecture has three key features. First, processing and memory cells are separated as two distinct function units for achieving easy data sharing and flexible memory usage. Second, a hierarchical NoC structure is adopted for providing high-bandwidth low-latency local communication and flexible global data routing. Third, in-cell configuration scheme employed in the cell array enables fast run-time context switching. To achieve a balanced design, three design methodologies, algorithm–architecture, hardware–software, and processing–memory co-design, are adopted during the construction of a reconfigurable cell array. Using the presented cell array as a baseline architecture, the following chapters present further developments and architectural improvements of the cell array, especially the processing and memory cells, through two case studies. Architectural developments are carried out by exploiting computational properties of the target application domain, namely digital baseband processing in wireless communication.

References

1. Z. Abdin, B. Svensson, Evolution in architectures and programming methodologies of coarse-grained reconfigurable computing. Microprocessors Microsyst. Embed. Hardw. Des. **33**, 161–178 (2009)
2. R. Airoldi, F. Garzia, O. Anjum, J. Nurmi, Homogeneous MPSoC as baseband signal processing engine for OFDM systems. in *International Symposium on System on Chip (SoC)*, Sept 2010, pp. 26–30
3. AMBA 4 AXI4-Stream Protocol Specification v1.0, Mar 2010
4. R. Baines, D. Pulley, A total cost approach to evaluating different reconfigurable architectures for baseband processing in wireless receivers. IEEE Commun. Mag. **41**(1), 105–113 (2003)

5. V. Baumgarte, G. Ehlers, F. May, A. Nückel, M. Vorbach, M. Weinhardt, PACT XPP-a self-reconfigurable data processing architecture. J. Supercomput. **26**, 167–184 (2003)
6. C. Bernard, F. Clermidy, A low-power VLIW processor for 3GPP-LTE complex numbers processing, in *Design, Automation Test in Europe Conference Exhibition (DATE)*, Mar 2011, pp. 1–6
7. T. Bjerregaard, S. Mahadevan, A survey of research and practices of network-on-chip. ACM Comput. Surv. **38**(1), 1 (2006)
8. B. Bougard, B. De Sutter, D. Verkest, L. Van der Perre, R. Lauwereins, A coarse-grained array accelerator for software-defined radio baseband processing. IEEE Micro **28**(4), 41–50 (2008)
9. J. Byrne, Tensilica DSP Targets LTE Advanced, Mar 2011. http://www.tensilica.com/uploads/pdf/MPR_BBE64.pdf
10. A. Chattopadhyay, Ingredients of adaptability: a survey of reconfigurable processors, in *VLSI Design*, Jan 2013
11. F. Clermidy, et al., A 477mW NoC-Based Digital Baseband for MIMO 4G SDR. in *IEEE International Solid-State Circuits Conference (ISSCC)*, Feb 2010, pp. 278–279
12. K. Compton, S. Hauck, Reconfigurable computing: a survey of systems and software. ACM Comput. Surv. **34**, 171–210 (2002)
13. M. Coppola, R. Locatelli, G. Maruccia, L. Pieralisi, A. Scandurra, Spidergon: a novel on-chip communication network, in *International Symposium on System-on-Chip*, 2004, p. 15
14. M. Dillinger, K. Madani, N. Alonistioti, *Software Defined Radio: Architectures, Systems and Functions*, 1st edn. (Wiley, New York, 2003)
15. A.Y. Dogan, J. Constantin, M. Ruggiero, A. Burg, D. Atienza, Multi-core architecture design for ultra-low-power wearable health monitoring systems, in *Design, Automation Test in Europe Conference Exhibition (DATE)*, 2012, pp. 988–993
16. J. Eker, J.W. Janneck, CAL language report: specification of the CAL actor language. Technical Report, University of California at Berkeley, Nov 2003
17. R. Fasthuber, et al., Exploration of Soft-Output MIMO detector implementations on Massive parallel processors. J. Signal Process. Syst. **64**, 75–92 (2011)
18. R. Hartenstein, A decade of reconfigurable computing: a visionary retrospective, in *Design, Automation Test in Europe Conference Exhibition (DATE)*, 2001, pp. 642–649
19. J. Janhunen, T. Pitkanen, O. Silven, M. Juntti, Fixed- and floating-point processor comparison for MIMO-OFDM detector. IEEE J. Sel. Top. Sign. Proces. **5**(8), 1588–1598 (2011)
20. G. Kahn, The semantics of a simple language for parallel programming, in *Information Processing* (North-Holland Publishing Company, Amsterdam, 1974), pp. 471–475
21. H. Lee, C. Chakrabarti, T. Mudge, A low-power DSP for wireless communications. IEEE Trans. Very Large Scale Integr. VLSI Syst. **18**(9), 1310–1322 (2010)
22. T. Lenart, Design of reconfigurable hardware architectures for real-time applications. Ph.D. thesis, Department of Electrical and Information Technology, Lund University, May 2008
23. Y. Lin, et al., SODA: a low-power architecture for software radio, in *International Symposium on Computer Architecture (ISCA)*, 2006, pp. 89–101
24. A. Nilsson, E. Tell, D. Liu, An 11 mm^2, 70 mW fully programmable baseband processor for mobile WiMAX and DVB-T/H in 0.12μm CMOS. IEEE J. Solid State Circuits **44**(1), 90–97 (2009)
25. A. Nilsson, Design of programmable multi-standard baseband processors. Ph.D. thesis, Department of Electrical Engineering, Linköping University, 2007
26. R.S. Patti, Three-dimensional integrated circuits and the future of system-on-chip designs. Proc. IEEE **94**(6), 1214–1224 (2006)
27. B. Plunkett, J. Watson, Adapt2400 ACM architecture overview. Quicksilver, 2004. A Technology White Paper
28. A. Rahimi, I. Loi, M. R. Kakoee, L. Benini, A fully-synthesizable single-cycle interconnection network for shared-L1 processor clusters, in *Design, Automation Test in Europe Conference Exhibition (DATE)*, 2011, pp. 1–6
29. M.A. Shami, A. Hemani, Morphable DPU: smart and efficient data path for signal processing applications, in *IEEE Workshop on Signal Processing Systems (SiPS)*, Oct 2009, pp. 167–172

30. M.A. Shami, A. Hemani, Classification of massively parallel computer architectures, in *IEEE 26th International Parallel and Distributed Processing Symposium Workshops PhD Forum (IPDPSW)*, May 2012, pp. 344–351
31. H. Svensson, T. Lenart, V. Öwall, Modelling and exploration of a reconfigurable array using systemC TLM, in *IEEE International Symposium on Parallel and Distributed Processing*, Apr 2008, pp. 1–8
32. H. Svensson, Reconfigurable architectures for embedded systems. Ph.D. thesis, Department of Electrical and Information Technology, Lund University, Oct 2008
33. M.B. Taylor, et al., A 16-issue multiple-program-counter microprocessor with point-to-point scalar operand network, in *IEEE International Solid-State Circuits Conference*, vol.1, Feb 2003, pp. 170–171
34. M. Thuresson, et al., FlexCore: utilizing exposed datapath control for efficient computing. J. Signal Process. Syst. **57**(1), 5–19 (2009)
35. T.J. Todman, G.A. Constantinides, S.J.E. Wilton, O. Mencer, W. Luk, P.Y.K. Cheung, Reconfigurable computing: architectures and design methods. Comput. Digit. Tech. **152**, 193–207 (2005)
36. K. van Berkel, F. Heinle, P.P.E. Meuwissen, K. Moerman, M. Weiss, Vector processing as an enabler for software-defined radio in handheld devices. EURASIP J. Appl. Signal Process. **2005**, 2613–2625 (2005)
37. Z. Yu, B.M. Baas, A low-area multi-link interconnect architecture for GALS chip multiprocessors. IEEE Trans. Very Large Scale Integr. VLSI Syst. **18**(5), 750–762 (2010)
38. C. Zhang, T. Lenart, H. Svensson, V. Öwall, Design of coarse-grained dynamically reconfigurable architecture for DSP applications, in *International Conference on Reconfigurable Computing and FPGAs (ReConFig)*, Dec 2009, pp. 338–343

Chapter 5
Multi-Standard Digital Front-End Processing

To demonstrate flexibility and performance of the reconfigurable cell array architecture introduced in Chap. 4, this chapter presents a case study of the platform configured for concurrent processing of multiple radio standards. Flexibility of the architecture is demonstrated by performing time synchronization and Carrier frequency offset (CFO) estimation for multiple Orthogonal frequency division multiplexing (OFDM)-based standards. As a proof-of-concept, this study focuses on three contemporarily widely used radio standards, 3GPP Long term evolution (LTE), IEEE 802.11n, and Digital video broadcasting for handheld (DVB-H). The employed reconfigurable cell array, containing 2×2 resource cells, supports all three standards and is capable of processing two concurrent data streams. The cell array is implemented in a 65 nm CMOS technology, resulting in an area of 0.48 mm^2 and a maximum clock frequency of 534 MHz. Dynamic configuration of the cell array enables run-time switching between different standards and allows adoption of different algorithms on the same platform. Taking advantage of the in-cell configuration scheme (described in Chap. 4), context switching between different operation scenarios requires at most 11 clock cycles. The implemented 2×2 cell array is fabricated as a part of a Digital front-end (DFE) Receiver and is measured as a standalone module via an on-chip serial debugging interface. Running at 10 MHz clock frequency and at 1.2 V supply voltage, the array reports a maximum power consumption of 2.19 mW during the processing of IEEE 802.11n data receptions and 2 mW during hardware configurations.

5.1 Introduction

Today, there is an increasing number of radio standards, each having different focuses on mobility and data rate transmission. For example, 3GPP LTE [3] aims to offer high mobility with moderate data rate; IEEE 802.11n [18] provides a high

© Springer International Publishing Switzerland 2016   49
C. Zhang et al., *Heterogeneous Reconfigurable Processors for Real-Time Baseband Processing*, DOI 10.1007/978-3-319-24004-6_5

data rate alternative for network services under stationary conditions; and DVB-H [12] specifically addresses multimedia broadcast services for portable devices. The evolution of radio standards continuously drives the development of underlying computational platforms with increased complexity demands. Meanwhile, requirements on time-to-market and Non-recurring engineering (NRE) cost force today's hardware platforms to be able to adopt succeeding amendments of standards. Doing modifications on dedicated hardware accelerators for each standard update is not affordable regarding both time and implementation cost. Furthermore, to obtain a continuous connection or constant data rate transmission, contemporary user terminals are expected to support more than one standard and to be able to switch between different networks at any time. Therefore, flexibility has become an essential design parameter to help computational platforms cope with various standards and support multiple tasks concurrently.

It is well identified that simultaneous support of multi-standard data receptions using flexible hardware platforms is a great challenge. Although some early attempts have been presented in both academia and industry [1, 7], experiments so far have been limited to the support of a single data stream. Switching between different standards is only possible through off-line configurations and is conducted by an external host controller. Despite not being reported, configuration time during context switching is envisioned to be on a scale of hundreds of clock cycles, since the host controller has to be interrupted to conduct the loading of appropriate programs/configurations before getting ready for new data receptions. Evidently, this off-line switching approach is undesirable from users' experience point of view, as terminals are temporarily "disconnected" every time when they enter a new radio environment. To address this issue, the European Union (EU) project "Scalable Multi-tasking Baseband for Mobile Communications" [10], or Multibase for short, is initiated. It aims to support concurrent processing of multiple data streams in a multi-standard environment and to provide seamless handover between different radio networks. The support of concurrent data streams improves user experiences, e.g., having simultaneous voice communication and video streaming. It also ensures continuous connectivity of user terminals, since an existing network connection can be maintained while a new network service is being established.

This study is carried out as a part of the EU Multibase project. The primary focus is on data computations in a DFE-Rx. As a proof-of-concept, three OFDM-based radio standards are selected: 3GPP LTE, IEEE 802.11n, and DVB-H. Besides, it is required to process two concurrent data streams from any of the three standards. Among processing tasks in a DFE-Rx, this study maps time synchronization and CFO estimation onto the reconfigurable cell array. Given that these tasks are performed during the (re)establishment of a data link between transmitter (Tx) and receiver (Rx), the required computational units are active only for a small fraction of the time when the receiver is on. This motivates the adoption of reconfigurable architectures in order to reuse hardware resources for other processing tasks. Hardware reusing can be exploited in two aspects. Firstly, in a *multi-standard single-stream* scenario, the same hardware can be reconfigured after OFDM synchronization to perform other baseband processing in succeeding stages, such as refined frequency

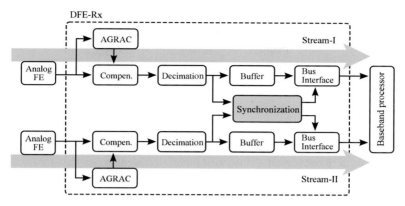

Fig. 5.1 Block diagram of the DFE-Rx constructed in the EU Multibase project [10]. This study focuses on the implementation of the synchronization block, *shaded* in the figure, by using the reconfigurable cell array

offset estimation and tracking. Secondly, in a *multi-standard multi-stream* scenario, underlying hardware resources can be shared for concurrent processing of multiple data streams. Figure 5.1 depicts the block diagram of the complete DFE-Rx and highlights the target processing block "synchronization." In the following, each function block of the DFE-Rx is briefly described. Implementation details can be found in [4, 6].

Taking digitized signals from the analog front-end, the Automatic Gain and Resource Activity Controller (AGRAC) and the compensation block adjusts the gain of incoming signals and performs DC-offset and IQ imbalance compensations, respectively. Besides, the AGRAC is used as a master core in the DFE-Rx to control the operation of other function blocks. The decimation filter chain contains a farrow resampler [8] used to adjust the sample rate of input data to that of the corresponding radio standards. Data resampling is one of the fundamental tasks when dealing with multiple radio standards, since the elementary sampling frequency varies between standards and is often not an integer multiple of one another. Since no pre-knowledge is given on which standard is going to be processed, the DFE-Rx has to operate at a frequency that is sufficiently high to capture signals of all standards without aliasing. Among the three standards under analysis, IEEE 802.11n has the highest sample rate, see Table 5.1, and is thus set as the master sampling frequency of the DFE-Rx. As a result, data streams of LTE and DVB-H after the decimation filter chain have an oversampling factor of $\lfloor 40/30.72 \rfloor^1 = 1$ and $\lfloor 40/9.14 \rfloor = 4$, respectively. The data reception buffer is used to store data inputs temporarily during front-end processing, and the bus interface block adapts the DFE-Rx to the following baseband processor.

---

[1] $\lfloor \cdot \rfloor$ is the floor function.

Table 5.1 Sampling frequency of the three target radio standards

	Bandwidth [MHz]	Number of subcarriers	Sampling frequency [MHz]
IEEE 802.11n [18]	40	128	40
3GPP LTE [3]	20	2048	30.72
DVB-H [12]	8	8192, 4096, 2048	9.14

The remainder of this chapter is organized as follows. Section 5.2 formulates the problem in more detail. Similarities and differences of the OFDM time synchronization in the target wireless radio standards are analyzed. Computational operations required by the synchronization process are elaborated. Section 5.3 presents hardware developments of the reconfigurable cell array with focuses on processing and memory cells. Section 5.4 starts by describing the computational and memory resource allocations during concurrent multi-standard processing. A software tool developed for generating hardware configurations of the cell array is presented. Implementation and silicon measurement results are summarized. The flexibility of the presented solution is demonstrated by mapping different algorithms onto the cell array after the chip is fabricated. Benefiting from the new algorithm mapping, the number of standards supported by the same cell array is further extended. Finally, Sect. 5.5 concludes this chapter.

5.2 Algorithm and Implementation Aspects

In OFDM, synchronization is needed due to the lack of common time and frequency references between Tx and Rx. An incorrect symbol timing may result in loss of orthogonality in the narrow-band subcarriers. Moreover, orthogonality may also be destroyed in the presence of a frequency mismatch between oscillators in Tx and Rx. OFDM synchronization makes sure of preserving orthogonality by providing a reliable start[2] of the OFDM symbol and CFO estimation.

The synchronization process is usually performed in time and frequency domain, commonly referred to as *acquisition* and *tracking* stage, respectively [13]. The acquisition stage aims to find the start of each OFDM symbol and to perform a rough estimation of CFO. The tracking stage aims to refine the parameters obtained from the acquisition stage. This study focuses on the acquisition stage and assumes that the channel impulse response is shorter than the length of CP.

---

[2]Orthogonality of narrow-band subcarriers is preserved as long as the estimated start lies within the Cyclic prefix (CP) of an OFDM symbol.

Fig. 5.2 IEEE 802.11n short and long training field [9]

5.2.1 Time Synchronization and CFO Estimation

Maximum-likelihood (ML) estimation [14] is commonly used to perform time synchronization in OFDM systems. The algorithm can be used on either pilots/preamble or CP. In the three standards under analysis, CP is present. Besides, IEEE 802.11n contains a preamble, which has specific Short training symbols (STSs) designed for data detection and time synchronization [18], see Fig. 5.2. Given that all STSs are identical, the first STS can be considered as the CP of the remaining part in the short training field (t_2–t_{10} in Fig. 5.2). Based on either CP or preamble, the ML estimation algorithm is expressed as

$$
\hat{\theta} =
\begin{cases}
\arg\max_{n}\{|\gamma[n]|\} & \text{if } |\gamma[n]| \geq T \\
\text{No symbol start found} & \text{otherwise}
\end{cases},
\tag{5.1}
$$

where

$$
\gamma[n] = \sum_{k=n-L+1}^{n} r[k]r^*[k-M].
\tag{5.2}
$$

In (5.1) and (5.2), $r[n]$ is the received data vector at sample index n, $\gamma[n]$ is the output of moving-sum, $\hat{\theta}$ indicates the estimated symbol start, and $(\cdot)^*$ denotes the complex conjugate operator. T represents the threshold value, which is used to find the symbol start by detecting the position of the maximum correlation value. T is a function of Signal-to-noise ratio (SNR) and is computed off-line and adjusted in accordance to different standards. L is the length of the moving-sum operation, and M is the autocorrelation distance, i.e., the number of samples from the start of CP to its corresponding copy within the OFDM symbol. The values of L and M vary among standards and also between different synchronization methods, CP-based for LTE and DVB-H and preamble-based for IEEE 802.11n. Table 5.2 summarizes the values of L, M, and the number of subcarriers N_c for the three standards. In CP-based synchronization, the autocorrelation distance M equals to N_c, and the size of the moving-sum L equals to the length of the CP. LTE and DVB-H fall into this category. Since better synchronization accuracy is expected when using preambles, preamble-based approach is used for IEEE 802.11n. In this case, M corresponds to the size of an STS and L equals to the size of remaining nine STSs ($L = 16 \times 9$ in Table 5.2). This is equivalent to computing correlation between neighboring STSs and accumulating results over the entire short training field.

Table 5.2 Comparison for the length of moving-sum L, auto-correlation distance M, and the number of subcarriers N_c in the ML-based time synchronization

	L	M	N_c
IEEE 802.11n	16×9	16	64
3GPP LTE	144	2048	2048
DVB-H	64	8192, 4096, 2048	8192, 4096, 2048

A common method to estimate CFO is to divide the offset value into two components, expressed as

$$\Delta f_c = \alpha + \varepsilon, \tag{5.3}$$

where α and ε represent the integer and fractional part of CFO, respectively. Both α and ε are normalized with respect to the subcarrier spacing. ε is delimited by $|\varepsilon| \leq 0.5$. This study focuses on the computation of the fractional CFO. An approach to estimate ε is based on a phase computation of the autocorrelation result at the estimated symbol start $\gamma[\hat{\theta}]$ [14], i.e.,

$$\varepsilon = \frac{1}{2\pi} \arg \left\{ \gamma[\hat{\theta}] \right\}. \tag{5.4}$$

This is usually performed by using a Coordinate rotation digital computer (CORDIC) algorithm operating in circular vectoring mode [11].

5.2.2 Operation Analysis

Based on the aforementioned algorithms, Fig. 5.3 shows a conceptual implementation diagram for performing OFDM acquisition. Operations are partitioned into three main processing blocks: *data correlation, peak detection, and CFO estimation.* Different design parameters in the three radio standards (Table 5.2) set different hardware requirements for the shaded blocks in Fig. 5.3. For example, the size of the correlation FIFO (M) changes from 16 to 2048 samples when switching from IEEE 802.11n to DVB-H 2K.

Input samples in this study are 12-bit complex numbers. To reduce memory requirements during the correlation computation, data samples in single-stream mode are truncated down to 4 bits. This relies on an assumption that performance of the synchronization in the acquisition stage only needs to be sufficiently accurate such that refined estimation algorithms in the tracking stage will work properly [5]. During concurrent multi-stream processing, memories are shared between two data streams and the wordlength of data samples is further reduced by half. As an example, performance analysis of CFO estimation with respect to different

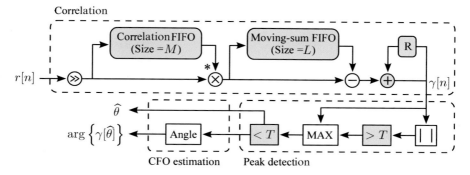

Fig. 5.3 Conceptual implementation diagram of the ML-based time synchronization and CFO estimation. *Shaded blocks* have various design parameters required by different radio standards

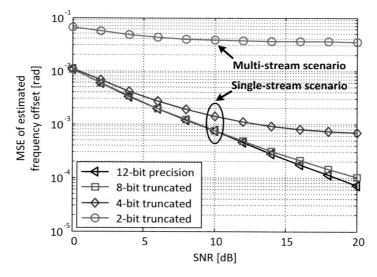

Fig. 5.4 Analysis of input data truncation in CFO estimation for 3GPP LTE with a frequency offset of $\pi/8$

input data truncation is shown in Fig. 5.4. The Mean squared error (MSE) of estimated frequency offset is simulated for an Additive white Gaussian noise (AWGN) channel on an LTE transmission with a frequency offset of $\pi/8$. Although higher data wordlength attains higher processing accuracy, larger input truncation reduces both hardware complexity and memory size. For example, reducing input wordlength from 12 to 4 bits results in a performance degradation of around 0.66×10^{-3} radians at an SNR of 10 dB. This corresponds to a frequency error of $0.66 \times 10^{-3}/(2\pi) \times \Delta f = 1.58$ Hz. Further truncation to 2 bits results in more than 80 % memory reduction at the cost of 87.26 Hz frequency error at the same SNR. Since a maximum frequency error of 2 kHz can be tolerated by the receiver in LTE [15], this wordlength reduction is motivated. The same analysis is applied to other radio standards, and results show that quantization noise due to wordlength reduction is negligible.

5.3   Hardware Development

Figure 5.5 shows a block diagram of the reconfigurable cell array deployed in the DFE-Rx for performing OFDM time synchronization and CFO estimation. Based on the operation analysis (Fig. 5.3), the cell array is configured to have two processing and two memory cells. The processing cells are used to perform data operations shown in Fig. 5.3, while the memory cells serve as correlation and moving-sum FIFOs as well as communication buffers between processors. The interface controller, connected to the cell array via hierarchical network interconnects, manages external data communication to other system blocks in the DFE-Rx and is responsible for static configurations of Resource cells (RCs). To cope with multi-standard concurrent data computations, both processing and memory cells in the baseline architecture described in Chap. 4 are further developed. The following sections present the detailed architectural improvements.

5.3.1   Dataflow Processor

In multi-standard multi-stream applications, concurrent processing calls for a processor design that suffices different computational requirements on each individual data stream. Processing cells employed in this study, named as dataflow processors, are Reduced instruction set computing (RISC) cores with improved dataflow control. In addition to the functions equipped in a generic processing cell (Chap. 4), the dataflow processor enhances data processing by supporting Single

Fig. 5.5   Block diagram of the 2 × 2 cell array and the interface controller deployed in the synchronization block of DFE-Rx. *Solid and dashed lines* depict local and hierarchical network interconnects, respectively. PC and MC denote for processing and memory cell, respectively

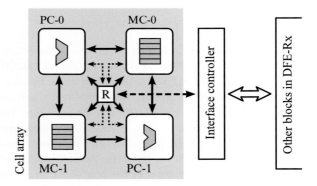

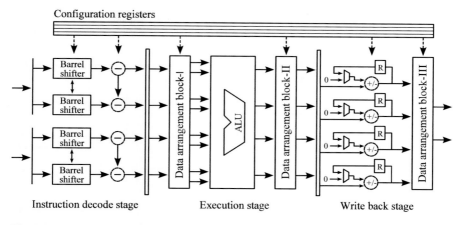

Fig. 5.6 Computation path of the dataflow processor. Configurations of the *shaded blocks* are stored in registers that are run-time accessible

instruction multiple data (SIMD)-like operations. The processor contains multiple processing lanes capable of performing both complex- and real-valued operations, illustrated in Fig. 5.6. These operations are required by, for example, ML estimation and CORDIC computations, respectively. Taking a 16-bit 4-lane processor as an example, the processing lanes can be grouped into two or four computation paths, capable of executing 8-bit complex-valued or 16-bit real-valued operations. Figure 5.7 depicts detailed architecture of the arithmetic part of Arithmetic logic unit (ALU) in the 16-bit processor. Basic operations of the ALU are controlled by two mode specifiers, "multiplication" and "vector." While the former one switches between addition and multiplication mode, the latter one controls real- or complex-valued operations. Real-valued output is obtained by concatenating results from 'O_3' and 'O_4', while the real and imaginary part of complex-valued output are taken from 'O_1' and 'O_4', respectively.

In addition, computational units are extended to both "instruction decode" and "write back" stage of the processor. As a result, several consecutive data manipulations can be accomplished in a single instruction execution without storing intermediate results. This substantially reduces register accesses. An example of the consecutive operation execution can be found during the computation of (5.2), where input data conjugate and result accumulation need to be performed before and after complex-valued multiplication, respectively. Without this extended computation capability, (5.2) needs to be computed iteratively, requiring three times more execution clock cycles and register accesses. Moreover, each arithmetic- and logic-type instruction in the dataflow processor is extended to have two operation codes (opcodes), which are capable of performing two different operations on the same input data operands in each clock cycle. The widely used butterfly operation (simultaneous add and subtract [2]) in Fast Fourier transform (FFT) is a typical example of the dual-opcode instruction. Another example is input data forwarding

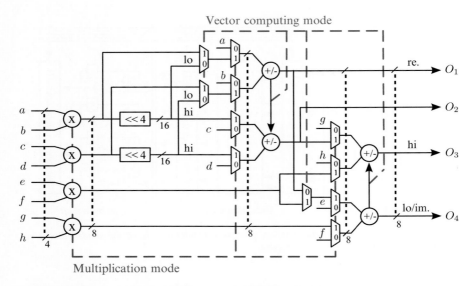

Fig. 5.7 Arithmetic part of the ALU in the dataflow processor, an example of 16-bit case. Real-valued output is taken from 'O_3' and 'O_4', while complex-valued output is drawn from 'O_1' and 'O_4'

during computations. This can be used to hide the execution time of data movement operations. For example, input data samples in (5.2) can be forwarded to other processing or memory cells while being multiplied and accumulated. The complete instruction set of the dataflow processor is included in Appendix A, Figs. A.1, A.2, A.3, and Tables A.1, A.2.

Data Stream Shuffling

For efficient usage of multiple processing lanes, the capability of redirecting input operands into either lane is vital. In the dataflow processor, data operands in each computation stage can be shuffled before and after each operation. Internal data shuffling is carried out by deploying data arrangement blocks at the computation stages, illustrated in Fig. 5.6. In this study, data arrangement blocks are built from multiplexers. Control bits of the multiplexers are stored in configuration registers that are transparently accessible by the user. Different data path configurations can be preloaded into the registers before executing a program or dynamically updated via a special instruction. The stored data path configurations can be applied to any type of instructions, which is accomplished by indexing the configuration registers in each instruction. As an example, trivial multiplications required in an FFT may be implicitly executed using data arrangement blocks, as the operations are equivalent to swapping two input operands without any data manipulation. Thus, specific operations to compute trivial multiplications are avoided, resulting

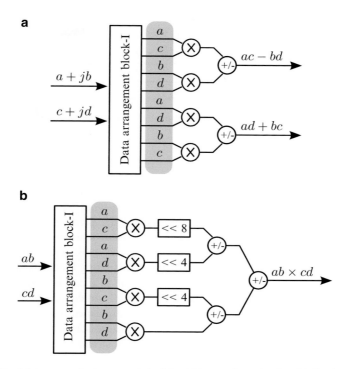

Fig. 5.8 Input data operand arrangements in (**a**) a 4-bit complex-valued multiplication and (**b**) an 8-bit real-valued multiplication. These two operations use the same instruction and data inputs but operate on different data sequences, as *highlighted* in the figure

in reduced execution clock cycles and program count. Moreover, with the help of data arrangement blocks, the same instruction can be used to perform different operations. This is accomplished by shuffling input operands to form different data patterns. As an example, real- and complex-valued multiplications share the same instruction but operate on two different data sequences, illustrated in Fig. 5.8. Detailed architecture of data arrangement blocks and the set of configuration codewords can be found in Appendix A, Fig. A.4, and Tables A.3, A.4, A.5, A.6.

5.3.2   Memory Cell

Under multi-standard multi-stream scenarios, memory descriptors are shared by the processing of multiple data streams. To cope with various sample rate of standards, it is crucial that memory descriptors can be executed in a non-sequential order. Otherwise, data stream with the slowest sample rate will block entire data processing.

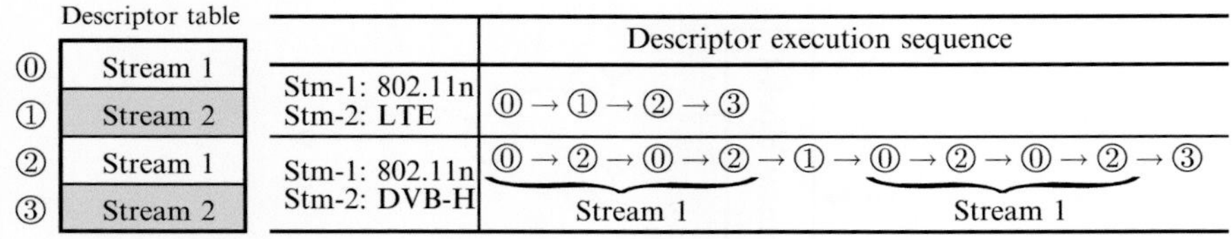

Descriptor table		Descriptor execution sequence	
⓪	Stream 1	Stm-1: 802.11n Stm-2: LTE	⓪ → ① → ② → ③
①	Stream 2		
②	Stream 1	Stm-1: 802.11n Stm-2: DVB-H	⓪ → ② → ⓪ → ② → ① → ⓪ → ② → ⓪ → ② → ③
③	Stream 2		Stream 1 — Stream 1

Fig. 5.9 Illustration of descriptor execution program during concurrent multi-stream processing

Flexible Descriptor Execution

Based on the aforementioned analysis, memory descriptors are extended in a way that they can be configured to execute either in *non-blocking* or *blocking* mode. In non-blocking mode, the operation controller of the memory cell sequentially starts a descriptor execution in each clock cycle, without waiting for response from data receiver regarding last memory access. Therefore, subsequent descriptors can still be issued and executed even if the current one is being blocked. Besides used in multi-stream processing, the non-blocking execution mode is also useful when one memory cell is shared among several hosts (e.g., processing cells) operating on different stream transfer rates. In contrast, blocking execution mode guarantees the completion of each specified memory access before starting a new descriptor execution. This mode can be used to avoid mixing up stream transfers when an I/O port is shared among multiple memory descriptors.

To further improve the flexibility of memory cells, the order of the descriptor execution is run-time programmable. This way, multiple descriptors can be arranged to reorder or repeat data sequences, or to cope with data streams that have different transfer rates. Figure 5.9 illustrates the use of descriptor execution program during concurrent multi-stream processing. Assuming that the memory cell has four descriptors, which are configured to serve for two different streams in an interleaved manner, namely ⓪ and ② for "stream 1" and ① and ③ for "stream 2". During the processing of IEEE 802.11n and LTE, which both have an oversampling rate of 1, the four descriptors are executed sequentially. However, when dealing with IEEE 802.11n and DVB-H, execution sequence needs to be programmed such that data stream of IEEE 802.11n is processed four times before performing one DVB-H data reception. This is due to the fact that data stream of DVB-H has an oversampling rate of 4, as mentioned in Sect. 5.1.

Micro-Block Function

In addition to the flexible descriptor execution, the data access pattern of a memory cell can be reshaped by using a micro-block function. This enables memory access with finer wordlength than that a physical memory provides. For example, a 32-bit wide memory cell can be configurable to behave as two 16-bit wide or four 8-bit wide memory cells. This feature is useful when supporting multi-standard

Table 5.3 Configuration fields of the memory descriptor

	Field	Bits	Description
Part I	dtype	31-30	Operation mode select
	base	29-20	Start address
	high	19-10	End address
	rd_ok/active	9	FIFO reading status/RAM active transfer flag
	wr_ok/rnw	8	FIFO writing status/RAM read–write select
	io_bank_rst	7	I/O port register reset
	id	6-1	Global packet destination ID
	block_opr	0	Blocking execution enable
Part II	rptr/ptr	31-22	Current FIFO reading pointer/RAM data pointer
	wptr/tsize	21-12	Current FIFO writing pointer/RAM data transfer size
	src/paddr	11-8	FIFO data source port/RAM address port
	dst/pdata	7-4	FIFO data destination port/RAM data port
	blk_size	3-1	Size of micro-block
	blk_en	0	Micro-block enable
Part III	blk_stride	31-27	Micro-block step size
	blk_rptr	26-22	Current micro-block read pointer
	blk_wptr	21-17	Current micro-block write pointer
	blk_mask	16-1	Micro-block data mask
	blk_mask_sign	0	Data mask sign extension

data processing, as different standards may intrinsically require different processing wordlength. Detailed usage of the micro-block function is further illustrated in Sect. 5.4.

A micro-block operation is defined by a *block size*, *stride*, *read and write pointer*, and *data mask*. The block size specifies the wordlength of a micro-block, used to determine the number of data accesses in each memory read and write operation. For a 32-bit memory cell, options of the micro-block size are 1, 2, 4, 8, 16, and 32 bits. Stride is the distance to the next micro-block, measured in bits. Read/write pointers are physical memory addresses and are automatically updated after each operation. Data mask enables bitwise operation on data that is read from or to be written to the memory. Table 5.3 summarizes configuration fields of the memory descriptor, which is an extended version of the one used in the baseline memory cell (Table 4.1 in Sect. 4.3.2).

5.4   Implementation Results and Discussion

Given that the cell array is aimed to be deployed in mobile terminals, a moderate clock frequency of around 300 MHz is expected. Considering the highest data sample rate among the three standards (40 MHz in IEEE 802.11n), operations assigned to each processing cell must be completed within eight clock cycles ($\lceil 300\,\text{MHz}/40\,\text{MHz}\rceil^3$) in order to suffice the requirement of real-time processing. In this study, concurrent processing of two data streams is accomplished by time-multiplexing two streams on the cell array. As a consequence, the required execution time is further reduced by half.

5.4.1 Task-Level Pipeline

To meet the stringent timing constraint, data computations are partitioned and mapped onto different processing cells. Specifically, the correlation and the peak detection in Fig. 5.3 are assigned to PC-0 and PC-1, respectively, while the CFO estimation is carried out on both processors after determining the position of the correlation peak. As mentioned in Sect. 5.2.1, the phase computation (5.4) in the CFO estimation can be performed using a CORDIC algorithm [11]. The concept of the CORDIC algorithm is to rotate input vector through a series of micro rotations by applying shift and add operations [19]. These operations can be easily mapped onto processing cells by using barrel shifters and ALU as well as data arrangement blocks for intermediate result shuffling. Memory cells interconnected with the two processors are used as correlation and moving-sum FIFOs. Additionally, MC-1 serves as a CORDIC coefficient Read-only memory (ROM) and communication buffers between the processors.

Based on the wordlength analysis in Sect. 5.2.2, the two processing cells are configured as 16-bit cores suitable for handling correlations with 4-bit complex-valued inputs (see Fig. 5.7). The wordlength of memory macros in both memory cells is 32 bits wide. Therefore, a pair of 16 bits in-phase and quadrature data inputs can be stored in the same memory location. Memory capacity of the correlation and the moving-sum FIFO is configured to suffice the standard with the largest storage requirement, i.e., DVB-H in this case.

5.4.2 Memory Interleaving

In addition to the task-level pipeline, self-governed FIFO operations in memory cells relieve processing cells from address manipulations. Moreover, the micro-block function of memory cells eliminates data alignment operations in processors, as illustrated in Fig. 5.10. Since the wordlength of memory macros is 32 bits wide and the complex-valued data inputs are truncated to 4 bits (in single-stream mode), four truncated data pairs need to be stored at one memory location. With the help of the micro-block function, PC-0 is exempted from data shifting and masking

---

$^3 \lceil \cdot \rceil$ is the ceiling function.

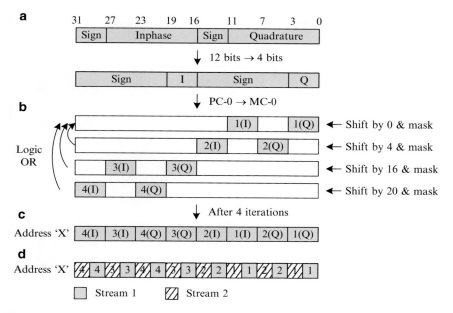

Fig. 5.10 Interleaved data storage in correlation FIFO, MC-0. (**a**) Received 12 bits data pair in PC-0. Data pairs are truncated down to 4 bits before being transmitted to MC-0. (**b**) Exploded view of data storage in correlation FIFO for a single data stream. (**c**) Final data storage at address 'X' in single-stream mode. (**d**) Final data storage at address 'X' for two concurrent data streams (2 bits data pair of each)

operations, as shown in Fig. 5.10b. The received data pairs are correspondingly left-shifted, masked to set off unused bits, and added (logic OR) to the write buffer. A final view of the interleaved data storage at a memory location in the correlation FIFO is illustrated in Fig. 5.10c. Similarly, when data pairs are truncated to 2 bits in the multi-stream mode, leading to eight micro-block operations, truncated data pairs from both streams are interleaved and stored in the same memory location, see Fig. 5.10d.

5.4.3 Context Switching

During context switching between different radio standards, both processing and memory cells need to be reconfigured to update parameters such as the threshold value T in the ML estimation (5.1), the length of the correlation (M), and the moving-sum FIFO (L). When switching from single- to multi-stream processing or vice versa, additional configurations are required. These include memory bank allocations, computational precision adjustments (e.g., from 4 bits down to 2 bits), and corresponding program segment updates in processing cells. Thanks

Fig. 5.11 Overview of the
configuration generation tool.
The memory cell MC-1 is
visualized as "MC 1" and
"MC 2" to ease
configurations in this study

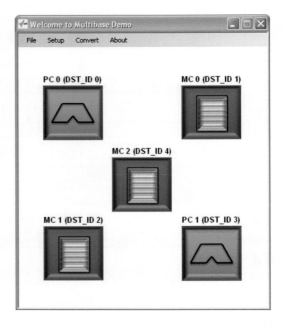

to the adoption of the in-cell configuration scheme presented in Chap. 4, all the
aforementioned context switching tasks are conducted inside the cell array during
system run-time. In the current implementation, PC-0 is used as a local master core
that controls the operations of other RCs and manages all the required resource
configurations.

5.4.4 Configuration Generator

To ease resource configurations of the cell array, a graphical user interface is
developed in-house, shown in Fig. 5.11. This tool visualizes all possible configurations of the processing and the memory cells deployed in the 2×2 cell array.
Additionally, it is able to convert resource configurations into binary codes and
generate a final bit stream based on allocated address space of each RC. Moreover,
this tool has the ability of streaming data and configurations into the cell array
(with the help of peripheral circuits, further discussed in Sect. 5.4.7) using a TCP/IP
socket. A detailed view of the tool can be found in Appendix A, Figs. A.5, A.6, and
A.7. Note that the memory cell MC-1 is shown as two separate units in Fig. 5.11,
"MC 1" and "MC 2", which are used as data memories (moving-sum FIFO and
CORDIC coefficient ROM) and communication buffers, respectively.

Table 5.4 Supported standards and memory utilization of a 2 × 2 cell array

Concurrency	Standards		Truncation wordlength	Memory utilization [%]
Single-stream	802.11n		4 bits	8.48
	LTE		4 bits	65.18
	DVB-H 2K		4 bits	85.71
	DVB-H 4K		2 or sign bit	85.71
	DVB-H 8K		Sign bit	85.71
Dual-stream	802.11n	& 802.11n	4 bits	16.96
	802.11n	& LTE	2 bits	45.09
	802.11n	& DVB-H 2K	2 bits	65.63
	LTE	& LTE	2 bits	73.21
	LTE	& DVB-H 2K	2 bits	93.75

5.4.5 Hardware Flexibility

With a fixed memory size, concurrent data processing is achieved by sharing memory resources between multiple data streams. Memory sharing is usually accomplished by sacrificing computational precision on all data streams, regardless of running standards and channel condition. Although the rigid uniform wordlength scheduling is easier to implement, higher computational precision is desired whenever possible. The cell array is capable of dynamically allocating computational resources to achieve better performance and resource utilization. Hence, computational precision is scheduled adaptively depending on the current operating condition. For the target radio standards, input samples are truncated to either 4 or 2 bits. Table 5.4 lists all the radio standards supported by the employed 2 × 2 cell array with the corresponding truncation wordlengths. Memory utilization, shown in the last column of Table 5.4, is the data memory used by the correlation process as a percentage of the total data storage available in all memory cells. The utilization does not reach 100 %, as parts of the data memories are used as the CORDIC coefficient ROM, as well as communication buffers between two processors.

In addition to resource sharing among multiple radio standards, system flexibility is also demonstrated by mapping different algorithms onto the same platform, without additional hardware cost. The adoption of different algorithms may either extend system compatibility by supporting additional standards or improve system performance by enhancing processing throughput and concurrency. In the conducted experiment, flexibility is illustrated by mapping a novel sign-bit OFDM synchronization algorithm [5] onto the presented cell array. This leads to the support of all three OFDM transmission modes (2K, 4K, and 8K subcarriers) in the DVB-H standard. The initial design choice on memory capacity only supports 4/2-bit synchronization algorithm for DVB-H 2K/4K modes. However, with the adoption of the sign-bit algorithm, which has dramatic data storage reduction,

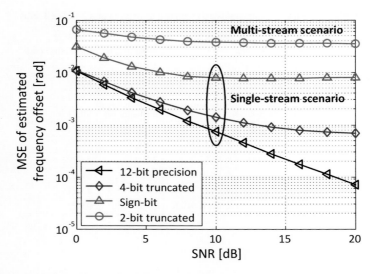

Fig. 5.12 Analysis of sign-bit algorithm [5] in CFO estimation for 3GPP LTE with a frequency offset of $\pi/8$

support for the DVB-H 8K mode becomes possible. Performance analysis of the sign-bit algorithm, see Fig. 5.12, reveals better computational accuracy than the 2-bit implementation. This is in virtue of specialized arithmetic used in the sign-bit algorithm, which obtains better immunity to quantization noise [15]. However, sign-bit implementation involves many bit-level data manipulations, which are difficult to map efficiently to a coarse-grained reconfigurable architecture without increasing execution time. Therefore, design complexity in the sign-bit algorithm shifts from memory capacity to data processing.

Figures 5.13 and 5.14 illustrate the layout of data storage in the correlation FIFO (MC-0) for various use cases adopted in this study. The considered use cases include different standards in the single-stream mode and various standard combinations in the multi-stream mode. Streams marked in red in Figs. 5.13 and 5.14 indicate the first data stream received by the cell array. Note that concurrent reception of IEEE 802.11n data streams can be processed in either 4 or 2 bits, because of low memory requirements, see Table 5.4.

5.4.6   Implementation Results

Fabricated in a 65 nm CMOS technology, the DFE-Rx has a die size of $5 \, \text{mm}^2$ with 144 I/O pads. According to synthesis results, half of the chip area is taken by memories, logic cells and I/O pads, while the remaining half is used for power and signal routing as well as clock tree generation. Figure 5.15 shows a chip layout. As a

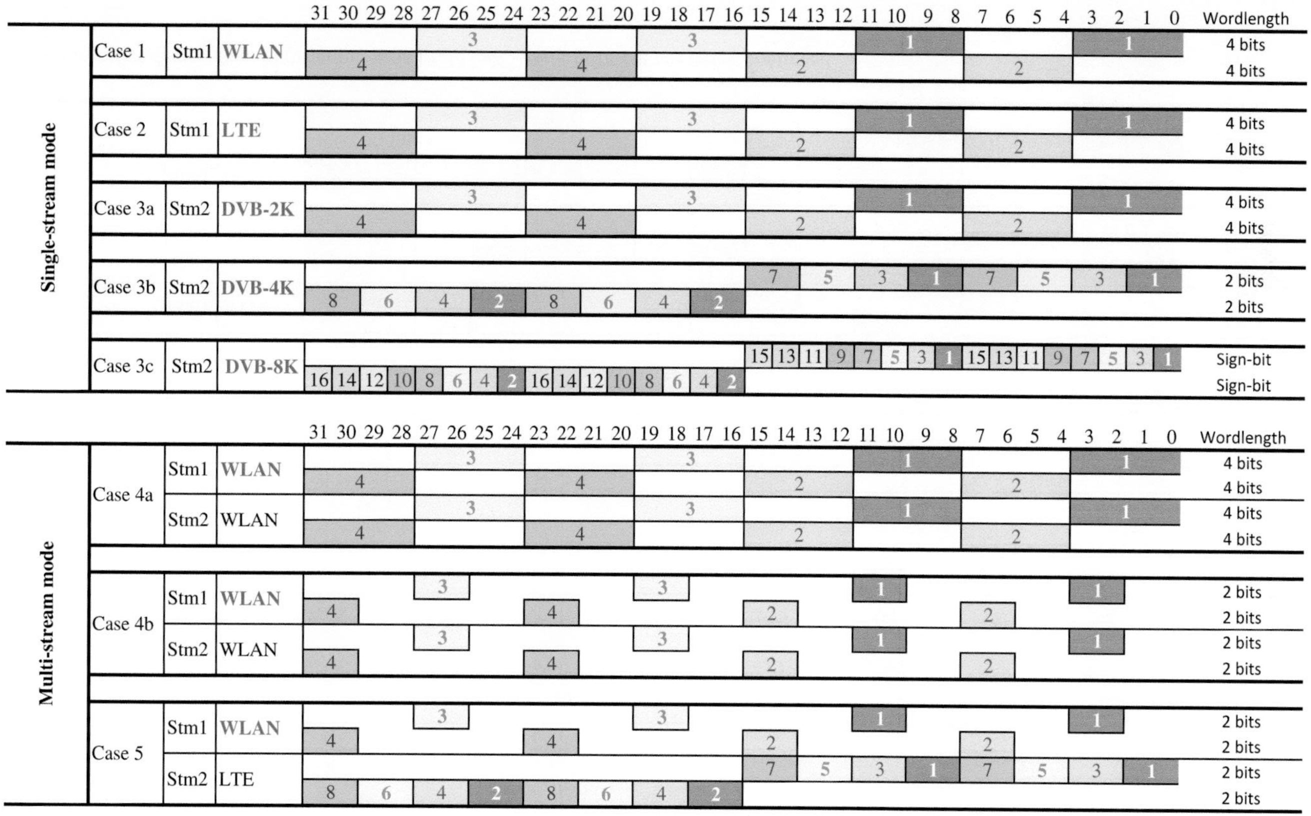

Fig. 5.13 Layout of data storage in the correlation FIFO (MC-0) for different use cases. Streams marked in *red* indicate the first data stream received by the cell array

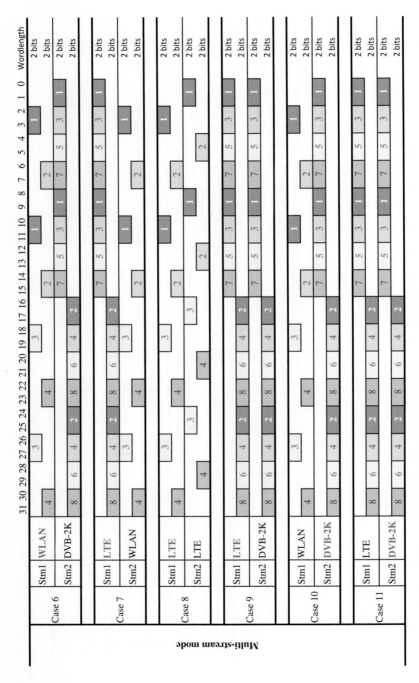

Fig. 5.14 Layout of data storage in the correlation FIFO (MC-0) for different use cases, continued

Fig. 5.15 Chip layout of the fabricated DFE-Rx

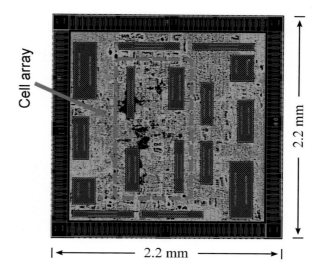

Cell array

2.2 mm

|← ———————— 2.2 mm ———————— →|

prototype, the design is pad-limited due to the large number of I/O ports required for individual tests of function blocks. Table 5.5 shows an area breakdown of the DFE-Rx. As can be seen, I/O pads occupy about 40 % of the area and the remaining part is evenly distributed among the synchronization block and the two receiving data paths. In the following, we focus on the implementation of the synchronization block, namely the 2×2 cell array, and present silicon measurement results obtained from a standalone test.

Shown by the synthesis results in Table 5.6, the cell array is memory dominant, which consumes about 40 % of the area. This is mainly due to the large amount of data required to store. Besides, program memories of processing cells are deployed such that they are large enough to allow further algorithm updates and mapping of other tasks. The entire cell array, including the interface controller, occupies $0.48 \, \text{mm}^2$ area and has a maximum clock frequency of 534 MHz. Thanks to the adopted in-cell configuration scheme, switching between different operation modes, such as from OFDM time synchronization to CFO estimation, requires only 11 clock cycles.

The high system flexibility offered by the cell array comes at the cost of area overhead. For comparison, a hardware accelerator only capable of performing the target tasks (OFDM time synchronization and CFO estimation) is implemented. Synthesis result of the accelerator is shown in Table 5.7. Even though a module-by-module comparison is not possible because of the hardware reusing nature of the reconfigurable architecture, it is evident that both designs are memory dominant. To be able to process two concurrent data streams, two accelerators are taken into the comparison. It shows that the accelerator-based solution uses around four times less silicon area and runs at a lower clock frequency (40 MHz) for the same throughput.

Table 5.5 Area breakdown
of the DFE-Rx

DFE-Rx	Area [μm^2]	Percentage [%]
I/O pads	991,569	38.37
Synchronization block	479,026	18.54
Rx data path 1	501,894	19.42
Rx data path 2	501,894	19.42
Others	109,851	4.25
Total	2,584,234	100.00

Table 5.6 Area breakdown of the 2×2 cell array in the DFE-Rx

Resource cell		Memory		Area [μm^2]	Percentage [%]
PC-0	Logic	—	—	37,567	7.84
	Memory	$384 \times 48b$	18 Kb	37,009	7.73
PC-1	Logic	—	—	37,201	7.77
	Memory	$512 \times 48b$	24 Kb	41,318	8.63
MC-0	Logic	—	—	39,167	8.18
	Memory	$512 \times 32b$	16 Kb	66,016	13.78
MC-1	Logic	—	—	64,810	13.53
	Memory	$384 \times 32b$	12 Kb	57,657	12.04
Router cells		—	—	37,976	7.93
Interface controller		—	—	60,306	12.59
Total			70 Kb	479,026	100.00

Table 5.7 Synthesis result of
a hardware accelerator, only
capable of performing time
synchronization and CFO
estimation for a single data
stream

	Area [μm^2]	Percentage [%]
Correlator	1296	2.13
Peak detector	2305	3.78
Correlation FIFO	27,232	44.69
Moving-sum FIFO	25,258	41.45
CORDIC (time-multiplexed)	3330	5.46
Control	1521	2.5
Total	60,942	100.00

However, from the entire DFE-Rx point of view, the adoption of the cell array results
in only about 16 % area overhead. In view of the high flexibility provided by the cell
array, as demonstrated in Sect. 5.4.5, this overhead is acceptable. Moreover, it should
be pointed out that the potential usage of the cell array is not fully explored when
only evaluating the mapping of the synchronization algorithms. The architecture
has the ability of being dynamically reconfigured to perform different tasks while a
hard-coded accelerator implements fixed functionality.

5.4.7 Measurement Results

To verify the functionality of the cell array, a standalone test is carried out in the debugging mode of the DFE-Rx via an on-chip Serial DeBuG (SDBG) interface. The SDBG interface, developed based on [16], contains a set of light-weight single-ended serial links capable of operating at 10 Mbps when using ribbon cable connections. Higher speed, up to 40 Mbps, can be achieved with good signal termination and PCB board layout.

Debugging Interface

Typical high speed data-recovery circuits require a Phase-Locked Loop (PLL) module for each serial link to recover data (as well as clock) from an 8b/10b encoded serial link. However, it has been shown in [16] that data can be recovered from a serial link by simply using a 4× clock sampling scheme without the 8b/10b encoding. Additionally, instead of using PLL, the presented data-recovery circuit employs two local clock signals: 'clk' and its 90-degree phase shifted counterpart 'clk90'. These clock signals can be generated from two independent local oscillators or clock generation circuits. However, it is important to maintain the phase-relationship of the two clocks. According to [16], this scheme can recover data from up to 200 Mbps serial links by using differential signaling for serial transmission and advanced Delay-Locked Loop (DLL) circuit to maintain the phase-relationship of clock signals. The current version of DFE-Rx has no differential I/O pads and DLL circuits equipped. Thereby, these serial links are implemented with single-ended I/O pads. Both clock signals ('clk' and 'clk90') are provided from normal clock pads and are directly used inside the chip without further phase adjustments.

The DFE-Rx SDBG consists of three serial links: ASIC Control Input Link (AIL-C), ASIC Data Input Link (AIL-D), and ASIC Output Link (AOL). The AIL-C and AIL-D are used to stream configurations and data into the cell array, respectively, while AOL is shared for both data and control outputs.

Standalone Cell Array Test

Through the SDBG interface, the cell array is connected to an FPGA platform, Xilinx XUPV5-LX110T, which implements the control and data streaming logics for communicating with the cell array. Figures 5.16 and 5.17 illustrate the setup and the measurement testbed for the standalone test, respectively.

For the system running in FPGA, a 32-bit MicroBlaze soft processor core [17] is used as a master controller. The SDBG interface adaptor is embedded as a co-processor connected on a shared Processor Local Bus (PLB). An interrupt controller is used to notify the master controller to receive control/data from the cell array. Communication to an external host (PC) is achieved through an Ethernet

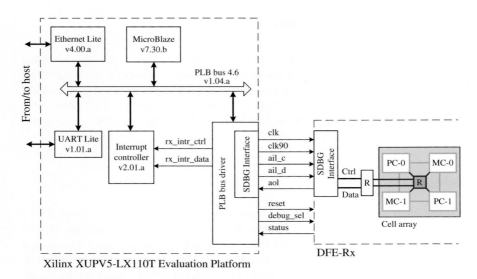

Fig. 5.16 Block diagram of the standalone cell array test setup

Fig. 5.17 Measurement testbed for the standalone cell array test

connection and a UART interface, which stream data/configuration and issue control commands to the cell array, respectively.

In the standalone test, the embedded system in FPGA operates at 100 MHz, whereas the cell array is clocked at 10 MHz due to the data rate limitations on ribbon cable connections and single-ended signaling. Data transmission rate over the Ethernet connection is set to 100 Mbps, while the UART line adopts a baud rate of 460.8 Kbps.

Table 5.8 Example of cell array configuration via a UART interface

≫ @g#	(Command input) Command 'g', destination RC selection
≫ 0	(Number input) RC hierarchical IO port ID
≫ @i#	(Command input) Command 'i', instruction downloading
≫ 2	(Number input) Number of instructions (plus one) to transfer
≫ $00010002#	(String input) Header of instruction downloading
≫ $A8000001#	(String input) Instruction "A8000001"

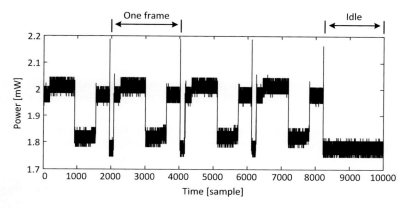

Fig. 5.18 Measured power consumption of the cell array in a standalone test when processing IEEE 802.11n data receptions

To provide users with an easy way of controlling the embedded system, a user interface in UART line is developed. Resource cell configurations and control/data inputs can be streamed into the cell array by issuing different user commands, see Appendix A, Table A.7. In addition, a few pre-loaded configuration scripts are provided for fast system demonstrations. Burst data transfers are accomplished by sending a user-defined script file. As an example, commands shown in Table 5.8 download an instruction into PC-0.

Based on the UART control, a high-level user interface is designed in MATLAB. The MATLAB interface provides a more advanced and flexible way to control data streams running into and out from the system. For example, input data sequences can be generated in MATLAB at run-time and data produced by the cell array can be collected and plotted graphically. Detailed user commands in the MATLAB interface are listed in Appendix A, Table A.8.

Figure 5.18 shows the power consumption of the cell array measured under the processing of an IEEE 802.11n data stream at nominal supply voltage of 1.2 V and at 10 MHz clock frequency. During the reception of 802.11n data frames, the measured minimum and maximum power consumption is 1.75 and 2.19 mW, respectively. During the loading of hardware configurations (not shown in Fig. 5.18), the cell array consumes 1.95–2 mW power.

5.5 Summary

This chapter presents a case study of the reconfigurable cell array suitable to process multiple radio standards concurrently. The flexibility and performance of the architecture are demonstrated by performing time synchronization and CFO estimation in OFDM systems. Using a 2×2 cell array, three widely used standards, IEEE 802.11n, 3GPP LTE, and DVB-H, are supported. Moreover, two independent data streams from the three standards can be processed concurrently by allocating and sharing system resources at run-time. During the reception of a single data stream, the cell array is configured to achieve high computational accuracy by using all available hardware resources. When two concurrent data streams are being received, the cell array adjusts its hardware resources such that data memories are split in two and processing cells are shared over time. Moreover, the potential of the architecture is further illustrated by mapping different algorithms onto the same platform without any additional hardware cost. Benefiting from the new algorithm mapping, the coverage of the standards supported by the cell array is extended. The employed 2×2 cell array is fabricated as a part of a Digital front-end (DFE) Receiver in the EU Multibase project. Running at 10 MHz clock frequency (in the standalone test) and at 1.2 V supply voltage, measurement results report a maximum power consumption of 2.19 mW during the processing of IEEE 802.11n data receptions.

References

1. A. Baschirotto, et al., Baseband analog front-end and digital back-end for reconfigurable multi-standard terminals. IEEE Circuits Syst. Mag. **6**(1), 8–28 (2006)
2. J.W. Cooley, J.W. Tukey, An algorithm for the machine calculation of complex Fourier series. IEEE J. Solid-State Circuits **19**(90), 297–301 (1965)
3. E. Dahlman, S. Parkvall, J. Sköld, 4G: LTE/LTE-Advanced for Mobile Broadband, 1st edn. (The Boulevard, Langford Lane, Kidlington, Oxford, UK, 2011)
4. I. Diaz, Algorithm-architecture co-design for digital front-ends in mobile receivers. Ph.D. thesis, Department of Electrical and Information Technology, Lund University, 2014
5. I. Diaz, L. Wilhelmsson, J. Rodrigues, J. Lofgren, T. Olsson, V. Öwall, A sign-bit auto-correlation architecture for fractional frequency offset estimation in OFDM, in *IEEE International Symposium on Circuits and Systems (ISCAS)*, May 2010, pp. 3765–3768
6. I. Diaz, et al., A new digital front-end for flexible reception in software defined radio. Microprocess. Microsyst. (2015). http://www.sciencedirect.com/science/article/pii/S0141933115000186
7. C. Ebeling, C. Fisher, G. Xing, M. Shen, H. Liu, Implementing an OFDM receiver on the RaPiD reconfigurable architecture. IEEE Trans. Comput. **53**(11), 1436–1448 (2004)
8. C.W. Farrow, A continuously variable digital delay element, in *IEEE International Symposium on Circuits and Systems*, vol. 3, Jun 1988, pp. 2641–2645
9. IEEE, IEEE P802.11N/D2.00. Technical report, IEEE LAN/MAN Standards Committee (Feb 2007)

10. Multi-base—scalable multi-tasking baseband for mobile communications (Feb 2008). ftp://ftp.cordis.europa.eu/pub/fp7/ict/docs/future-networks/projects-multibase-factsheet-20080206_en.pdf
11. B. Parhami, Computer Arithmetic: Algorithm and Hardware Designs (Oxford University Press, Oxford, New York, 2000)
12. U.H. Reimers, DVB-the family of international standards for digital video broadcasting. Proc. IEEE **94**(1), 173–182 (2006)
13. M. Speth, S. Fechtel, G. Fock, H. Meyr, Optimum receiver design for OFDM-based broadband transmission – Part II: a case study. IEEE Trans. Commun. **49**(4), 571–578 (2001)
14. J.J. van de Beek, M. Sandell, P.O. Borjesson, ML estimation of time and frequency offset in OFDM systems. IEEE Trans. Signal Process. **45**(7), 1800–1805 (1997)
15. L. Wilhelmsson, I. Diaz, T. Olsson, V. Öwall, Performance analysis of sign-based pre-FFT synchronization in OFDM systems, in *IEEE 71st Vehicular Technology Conference*, May 2010, pp. 1–5
16. Xilinx, Application Note XAPP224: Data Recovery (Jul 2005)
17. Xilinx, MicroBlaze Processor Reference Guide (v14.1) (Apr 2012)
18. X. Yang, IEEE 802.11n: enhancements for higher throughput in wireless LANs. IEEE Wirel. Commun. **12**(6), 82–91 (2005)
19. C. Zhang, T. Lenart, H. Svensson, V. Öwall, Design of coarse-grained dynamically reconfigurable architecture for DSP applications, in *International Conference on Reconfigurable Computing and FPGAs (ReConFig)*, Dec 2009, pp. 338–343

Chapter 6
Multi-Task MIMO Signal Processing

Driven by the requirement of multi-dimensional computing in contemporary wireless communication technologies, reconfigurable platforms have come to the era of vector-based architectures. In this chapter, the reconfigurable cell array developed in Chaps. 4 and 5 is extended with extensive vector computing capabilities, aiming for high-throughput baseband processing in MIMO-OFDM systems. Besides the heterogeneous and hierarchical resource deployments, a vector-enhanced SIMD structure and various memory access schemes are employed. These architectural enhancements are designed to suffice stringent computational requirements while retaining high flexibility and hardware efficiency. Implemented in a 65 nm CMOS technology, the cell array occupies 8.88 mm^2 core area. To illustrate its performance and flexibility, three computationally intensive blocks, namely channel estimation, channel matrix pre-processing, and symbol detection, of a 4×4 MIMO processing chain in a 20 MHz 64-QAM Long term evolution-advanced (LTE-A) downlink are mapped and processed in real-time. Operating at 500 MHz and 1.2 V voltage supply, the achieved throughput is 367.88 Mb/s and the average power consumption is 548.78 mW. The corresponding energy consumption for processing one information bit is 1.49 nJ. Comparing to state-of-the-art implementations, the presented solution outperforms related programmable platforms by several orders of magnitude in energy efficiency, and achieves similar level of area and energy efficiency to that of ASICs.

6.1 Introduction

Given data receptions processed in DFE-Rx (Chap. 5) and transformed back to the frequency domain via FFT [54], this chapter focuses on succeeding blocks in the baseband processing chain of the receiver and considers systems employing MIMO and OFDM technologies. In addition, 3GPP LTE-A is used as a case study to

© Springer International Publishing Switzerland 2016
C. Zhang et al., *Heterogeneous Reconfigurable Processors for Real-Time Baseband Processing*, DOI 10.1007/978-3-319-24004-6_6

illustrate architectural development of the reconfigurable cell array. Support for other or multiple radio standards can be developed using the similar approach. However, the focus of this study is on vector enhancements of the cell array and concurrent processing of multiple tasks.

Compared to single antenna systems, MIMO technology exploits design-of-freedom in the spatial domain in addition to time and frequency. Therefore, it provides significant improvements in system capacity and link reliability without increasing bandwidth. However, the price-to-pay is higher complexity and energy consumption due to increased computational dimensions, i.e., in proportion to the antenna size, and required sophisticated signal processing, such as symbol detection for inter-antenna interference cancellations. Moreover, when combining MIMO with OFDM, it is required to perform the corresponding processing at every OFDM subcarrier, posing even more stringent computational and energy requirements. Under such circumstances, building architectures solely on scalar-based function units requires a large number of resource deployments. Although the adoption of massive scalar units provides high flexibility, it reveals poor hardware efficiency in view of the parallel-structured MIMO-OFDM processing, since a large portion of resource controls (e.g., processor instructions and memory configurations) are identical and thus redundant. To achieve efficient computing, it is crucial to extend architectures with vector processing capabilities by fully utilizing the extensive Data-level parallelism (DLP) available in MIMO-OFDM systems.

In this chapter, the reconfigurable cell array presented in Chaps. 4 and 5 is further developed, aiming at achieving a balance between processing performance, flexibility, and hardware efficiency. Specifically, the architecture is partitioned into distinct vector and scalar processing domains for efficient hybrid-format data computing. In the vector domain, processing cells are deployed with vector-enhanced SIMD cores and VLIW-style multi-stage computation chains to attain low-latency high-throughput vector computing [55]. Memory cells are equipped with flexible vector access schemes for relieving non-computational address manipulations from processing cells. For performance evaluation, three tightly coupled baseband processing blocks, which are unique and crucial in MIMO for exploiting its full superiorities, are mapped onto the cell array in a time-multiplexed manner. The three processing blocks are:

- *Estimation* of the channel state information using pilot tones,
- *Channel matrix pre-processing* that is an indispensable step for all kinds of detection algorithms,
- *Symbol detection* that recovers the transmitted vector.

Figure 6.1 shows a simplified diagram of the MIMO-OFDM transceiver and highlights the target processing blocks in this study. In addition to the vector extension and task-level multiplexing, hardware efficiency of the cell array is further improved by algorithm-level exploitation, in which more than 98 % of the total operations involved in all three tasks are vectorized and unified, enabling extensive parallel processing and hardware reuse.

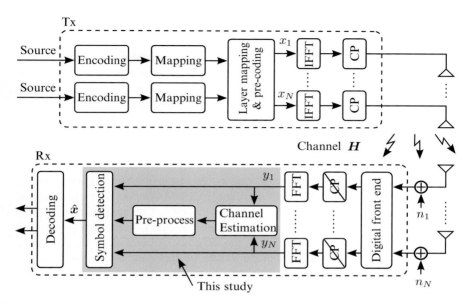

Fig. 6.1 Block diagram of the MIMO-OFDM transceiver. This study maps all three *shaded blocks* onto the reconfigurable cell array

In Sect. 6.2, algorithm development for the MIMO processing tasks are presented, including Minimum mean-square error (MMSE)-based channel estimation, QR decomposition (QRD)-based channel matrix pre-processing, and node-perturbation-enhanced MMSE symbol detection. In Sect. 6.3, performance, computational complexity, and hardware friendliness of the adopted algorithms are evaluated and analyzed in comparison with conventional approaches. Based on the operation profile and the LTE-A specification, data processing flow and timing analysis are conducted to provide guidance for succeeding hardware development. Section 6.4 describes the detailed array architecture configured for MIMO-OFDM signal processing. The focus is on vector extension, including heterogeneous resource arrangement, processing cell enhancements, and various vector memory access schemes. Section 6.5 summarizes implementation results and compares performance with that of state-of-the-art platforms. Moreover, the flexibility of the presented solution is further illustrated in Sect. 6.6 through a mapping of an adaptive channel matrix pre-processor. The mapping makes use of dynamic resource allocations to adopt appropriate pre-processing algorithms at run-time, which provides a wide range of performance-complexity trade-offs. Section 6.7 concludes this chapter.

6.2   MIMO Signal Processing

This study considers a MIMO-OFDM system with N transmit and receive antennas. Without loss of generality, the following discussions are based on the consideration of 20 MHz LTE-A downlink operating in normal Cyclic prefix (CP) mode with 4×4 antenna setup and 64-QAM modulation. Similar to the previous chapter, the maximum excess delay of the propagation channel is assumed to be within the CP of each OFDM symbol. Before presenting the target MIMO processing algorithms, the LTE-A data structure is introduced and the system model described in Chap. 3 is briefly revisited.

Figure 6.2 shows the structure of a resource block in LTE-A. Each resource block contains 12 consecutive subcarriers and 7 OFDM symbols over a time slot of 0.5 ms. To support operations such as synchronization and channel estimation, pilot tones are distributed over the time-frequency grid. Under the common assumption of quasi-static[1] channel modeling [23], the accuracy of channel estimates can be improved by inserting pilot tones in the middle of each time slot (symbol 4) into the pilot vector at symbol 0 [45]. This way, channel estimation is performed only once in each time slot, followed by channel matrix pre-processing, while symbol detection is required on every data-carrying subcarrier.

Assuming perfect synchronization and front-end processing, the received vector y after CP removal and FFT can be expressed as

$$y = Hx + n, \tag{6.1}$$

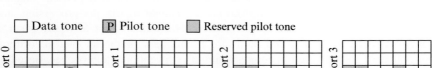

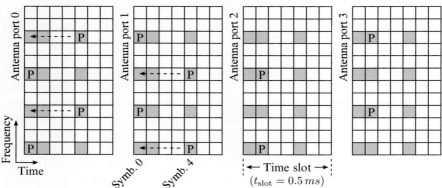

Fig. 6.2 Pilot pattern for four antenna ports in one LTE-A resource block

---

[1]Channel coefficients of each subcarrier are stationary over time within one time slot, i.e., 0.5 ms in LTE-A.

where H denotes the complex-valued channel matrix, x is the transmitted vector obtained by mapping a set of encoded information bits onto a Gray-labeled complex constellation, and n is the Independent and identically distributed (i.i.d.) complex Gaussian noise vector with zero mean and variance σ_n^2. The average transmit power of each antenna is normalized to one, such that $E\{xx^H\} = I_N$, where I_N is an identity matrix of size N and $(\cdot)^H$ denotes a Hermitian transpose.

In the following, algorithms adopted for the three processing tasks are presented in detail. It should be pointed out that algorithm selections are not the main scope of this chapter. They are selected to make use of essential architectural characteristics and to illustrate the performance of the hardware platform.

6.2.1 Channel Estimation

Based on the scattered pilot arrangement (Fig. 6.2), pilot-aided comb-type channel estimation scheme [25] is employed. It consists of two computation steps. First, channel coefficients at pilot positions (denoted by subscript p) are computed by using Least square (LS) algorithm,

$$h_{p,\text{LS}} = y_p x_p^{-1}. \tag{6.2}$$

Second, channel coefficients at data-carrying subcarriers are estimated by interpolating and extrapolating $h_{p,\text{LS}}$ in the frequency domain. This process may be seen as a linear filtering of the LS estimation

$$h = \mathcal{W} h_{p,\text{LS}}, \tag{6.3}$$

where the filter function $\mathcal{W}$ varies with different algorithms (mentioned in Chap. 3.2). Among them, Linear MMSE (LMMSE) estimation algorithm aims to approach the optimal result by using second-order statistics of the channel conditions and noise power. It is defined as

$$h_{\text{MMSE}} = \mathcal{W} h_{p,\text{LS}} = R_{h_d h_p} \left(R_{h_p h_p} + \frac{\beta}{\text{SNR}} I_N \right)^{-1} h_{p,\text{LS}}, \tag{6.4}$$

where $R_{h_d h_p}$ is the channel cross-correlation between pilot and data-carrying subcarriers, $R_{h_p h_p}$ represents the channel auto-correlation between pilot subcarriers, SNR denotes the average signal-to-noise ratio of received signals, and β is a constellation dependent constant, e.g., $\frac{180}{67}$ for 64-QAM.

Robust MMSE Estimator

A major drawback of the LMMSE estimator is its high computational complexity.
One reason for this is the need for re-computation of W every time SNR and/or
the correlation matrices change. This is infeasible for practical implementations,
especially when the number of subcarriers is large. Instead, the Robust MMSE
(R.MMSE) algorithm [14] is adopted in this study. The R.MMSE estimator
completely removes the need for run-time W calculations by employing a static
function, designed to safely tolerate various channel scenarios and rapid channel
variations. The static W is generated by using underestimated correlation matrices
and an overestimated SNR value. Specifically, the correlation matrices are pre-
computed based on two assumptions. First, the propagation channel obeys a uniform
PDP. Second, the maximum excess delay of the channel is equal to the length
of CP. This is illustrated in Fig. 6.3a. For comparisons, PDPs defined for some
typical channel scenarios in LTE-A are included. As shown, they are all covered
by the envelope of the uniform PDP used in R.MMSE, revealing the underestimated
design strategy. Regarding SNR, a high value is preferable. This can intuitively be
explained by considering a high-pass filtering process, where the value of SNR acts
as the cut-off frequency of the filter. Channel estimation errors are concealed in
noise with low SNR (i.e., being attenuated in the stop-band region), but tend to
be pronounced with high SNR. Hence, it is better to push SNR towards a high
value region to keep the channel estimation error low for a large SNR range. Using
these static parameters, W becomes a constant scaling matrix that may be prepared
off-line. To sum up, compared to LMMSE, the R.MMSE approach reduces the
complexity at the cost of increased estimation error. Nevertheless, it performs better
than the LS estimator, because of the use of second-order channel statistics and the
reduction of noise enhancements.

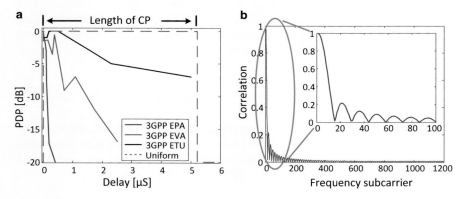

Fig. 6.3 (**a**) The uniform PDP used in the R.MMSE estimator. PDPs of some typical channel
scenarios in LTE-A [1] are included for comparisons. (**b**) An illustration of the frequency-domain
correlation between subcarriers in 20 MHz LTE-A, under the case of uniform PDP

Modified Robust MMSE Estimator

Although (6.4) in R.MMSE is reduced to a constant matrix multiplication, it is still computationally intensive when considering the dimension of $h_{p,\text{LS}}$. For example, the vector has a size of 400×1 for 20 MHz LTE-A. To further reduce the complexity, a sliding window approach is applied to the R.MMSE algorithm, named as R.MMSE-SW for short. The key is to apply low-rank approximations [14] on $R_{h_d h_p}$ and $R_{h_p h_p}$ based on the fact that adjacent subcarriers generally have dominant contribution to correlation coefficients. Figure 6.3b illustrates the correlation between the first frequency subcarrier and all remaining ones under the case of the uniform PDP. As an example, the first 100 subcarriers contribute to more than 95 % of the total correlation value. Therefore, a frequency correlation window (N_{SW}) containing only a number of neighboring pilots is applied to each estimation. As a consequence, the size of the matrix multiplication in (6.4) is dramatically reduced compared to the full-window case, i.e., when the correlation of all subcarriers are considered. In R.MMSE-SW, the size of N_{SW} is a performance-complexity trade-off parameter, which may be adjusted depending on the channel condition and performance demand.

6.2.2 Channel Matrix Pre-processing

Each estimated channel matrix $\hat{H}$ needs to be further processed before being sent to the succeeding symbol detector. There are two commonly used channel matrix pre-processing methods, inversion and QRD of the channel matrix, required in linear and tree-search based detectors, respectively. For matrices of size 4×4 or larger, it has been shown in [15] that matrix inversion can be efficiently computed by means of QRD. Hence, the QRD-based channel matrix pre-processing method is adopted. In addition, sorting is applied to the channel matrix during the QRD process, aiming to improve the detection performance. This is commonly referred to as Sorted QR decomposition (SQRD) [49].

SQRD starts by column-wise permuting $\hat{H}$ based on the post-detection SNR [24] of each spatial stream. Thereafter, $\hat{H}$ is decomposed into an unitary matrix Q and an upper triangular matrix R with real-valued non-negative elements on the main diagonal. With these, the QRD-based channel matrix pre-processing is expressed as

$$\hat{H} = \hat{H}_p P^T = QRP^T, \tag{6.5}$$

where the $N \times N$-dimensional permutation matrix P contains the corresponding sorting sequence and the operator $(\cdot)^T$ denotes a matrix transpose.

MMSE-SQRD Algorithm

Considering the use of MMSE criterion during the adopted symbol detection, presented further in Sect. 6.2.3, MMSE-SQRD algorithm [50] is employed to pre-process $\hat{H}$. The basic idea is to reduce the probability of ill-conditioned channel matrix by taking the additive noise into account. This is equivalent to performing SQRD of an augmented matrix of size $2N \times N$,

$$\underline{\hat{H}} = \begin{bmatrix} \hat{H} \\ \sigma_n I_N \end{bmatrix} = \underline{\hat{H}}_p P^T = \underline{QR} P^T = \begin{bmatrix} Q_a \\ Q_b \end{bmatrix} \underline{R} P^T, \tag{6.6}$$

where Q_a, Q_b, and $\underline{R}$ have the same size as $\hat{H}$, i.e., $N \times N$. One interesting property of (6.6) is that $\underline{R}^{-1}$ is obtained as a by-product of the matrix decomposition process, i.e.,

$$\underline{R}^{-1} = \frac{1}{\sigma_n} Q_b. \tag{6.7}$$

Therefore, no explicit matrix inversion is needed when computing $\underline{\hat{H}}^{\dagger}$, where $(\cdot)^{\dagger}$ denotes a matrix pseudo-inverse. With the augmented matrix $\underline{\hat{H}}$, the system model in (6.1) can be rewritten as

$$\tilde{y} = \underline{R} x_p + \tilde{n}_p, \tag{6.8}$$

where

$$\tilde{y} = \underline{Q}^H \begin{bmatrix} y \\ 0_{N \times 1} \end{bmatrix} = Q_a^H y, \tag{6.9}$$

$x_p = P^T x$ is the row-wise permuted x, and $\tilde{n}_p = Q_a^H n$ is the noise vector that has the same statistics as n.

Compared to (6.5), the decomposition of $\underline{\hat{H}}$ results in an increased computational complexity, by roughly 50 %, because of the doubled matrix dimension. However, MMSE-SQRD improves the performance of linear detectors and achieves a significant complexity reduction in tree-search based detection algorithms [36].

Iterative Sorting and MGS-QRD Algorithm

The sorting of $\hat{H}$ involves a matrix inversion, required for evaluating the post-detection SNR (η_i) of each spatial stream. Under the assumption of $E\{xx^H\} = I_N$, η_i is defined as [24]

$$\eta_i = \frac{1}{\sigma_n^2 \left(\hat{\underline{H}}^H \hat{\underline{H}} \right)^{-1}_{i,i}}, \qquad (6.10)$$

where $(\cdot)_i$ denotes the column vector and $(\cdot)_{i,i}$ is the (i,i)th matrix element. For practical implementations, approximation of (6.10) is commonly used for reducing the computational complexity, such as the one suggested in [49]

$$\eta_i \sim \left\| \hat{\underline{h}}_i \right\|_2^2, \qquad (6.11)$$

with $\| \cdot \|_2$ denoting ℓ^2-norm. Based on (6.11), various sorting strategies exist. Commonly used ones are one-time and iterative sorting [49]. Different from the former one, the iterative sorting approach keeps track of column changes in $\hat{\underline{H}}$ during the decomposition process, which leads to a better sorting result. The iterative sorting steps are shown in Algorithm 1 on lines 3, 5, and 12.

For QRD computations, several well-known methods exist, such as Gram-Schmidt orthogonalization, Householder transformation, Givens rotation, and their derivatives [19]. The Gram-Schmidt process obtains the orthogonal basis spanning the column space of the matrix by the orthogonality principle [27]. The Householder transformation handles column vectors of the matrix by reflection operations [19]. Givens rotation operates on one element at a time by using a sequence of unitary transformations [36]. Considering the accuracy and numerical stability, computational complexity, and hardware reusability, Modified Gram-Schmidt (MGS) [19] method is used for implementing the QRD in this study. The MGS-QRD algorithm iteratively computes the $\underline{Q}$ and the $\underline{R}$ matrix in N steps. Core operations of MGS-QRD per iteration i are summarized in Algorithm 1 from lines 7 to 12.

Algorithm 1 MGS-based MMSE-SQRD algorithm

1: $\hat{\underline{H}} = [\hat{H}, \sigma_n I_N]^T$
2: $\underline{Q} = \hat{\underline{H}}; \quad \underline{R} = 0_{N \times N}; \quad P = I_N$
3: $\xi = \left[\| \underline{q}_1 \|_2^2, \| \underline{q}_2 \|_2^2, \ldots, \| \underline{q}_N \|_2^2 \right]^T$
4: **for** $i = 1, 2, \ldots, N$ **do**
5:     $j = \arg \min_{l=i,i+1,\ldots,N} \xi_l$ **% Iterative sorting**
6:     Exchange columns/elements i and j in $\underline{Q}, \underline{R}, P$, and ξ
7:     $\underline{r}_{i,i} = \sqrt{\xi_i}$
8:     $\underline{q}_i = \underline{q}_i / \underline{r}_{i,i}$
9:     **for** $k = i + 1, i + 2, \ldots, N$ **do**
10:         $\underline{r}_{i,k} = \underline{q}_i^H \underline{q}_k$
11:         $\underline{q}_k = \underline{q}_k - \underline{r}_{i,k} \underline{q}_i$
12:         $\xi_k = \| \underline{q}_k \|_2^2$ **% Column-norm updating**
13:     **end for**
14: **end for**

6.2.3   Symbol Detection

With the received signal y and pre-processed channel matrix $\hat{\underline{H}}$, transmitted vector x is recovered by using a MIMO symbol detector. As described in Chap. 3.2, there are two commonly used detection schemes. Linear detection algorithms, such as Zero-forcing (ZF) and MMSE, mainly consist of vector operations and thus are architecture-friendly to vector-based platforms. However, they suffer from significant performance degradation compared to the optimal Maximum-likelihood (ML) detection, especially in frequency-selective fading channels. On the other hand, near-ML tree-search algorithms, e.g., Sphere decoder (SD), K-Best, and their derivatives [7, 22], do not map efficiently to vector-based architectures. This comes from the fact that the tree-search procedure deals with one spatial layer at a time and involves massive sequential scalar operations, which are frequently switched between node expansion, partial Euclidean distance sorting, and branch pruning. Therefore, the native vector structure of MIMO data streams is destroyed. To tackle this problem, detection algorithms, such as Fixed-complexity sphere decoder (FSD) [4] and Selective Spanning with Fast Enumeration (SSFE) [37], are used to bring in vectorized operations that may be performed independently at each layer. However, they do not solve the essential problem of tree-structured detection schemes. Data dependency between spatial layers still restrains the full potential of parallel architectures.

To bridge the algorithm–architecture gap, illustrated in Fig. 6.4, a highly parallelized symbol detection algorithm is developed. It provides near-ML performance, like tree-search algorithms, while retaining the inherent vectorized operations of linear detection schemes. This is achieved by employing a vector-level closest point search scheme in conjunction with linear detectors. In this study, the adopted algorithm is built upon a linear MMSE detector and is named as Node-Perturbation-enhanced MMSE (MMSE-NP). As a proof-of-concept, this study focuses on the hard-output symbol detection.

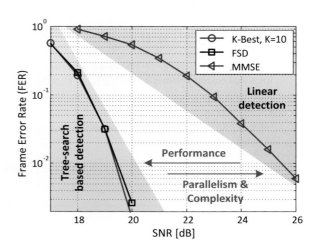

Fig. 6.4 An illustration of the algorithm–architecture gap between linear and tree-search based detection schemes

Parallel Node Perturbation

MMSE-NP starts by obtaining an initial result using a linear MMSE detection

$$x_p^{\text{MMSE}} = \underline{R}^{-1}\tilde{y} = \frac{1}{\sigma_n}Q_bQ_a^H y. \tag{6.12}$$

Hard-output detection result $\hat{x}_p^{\text{MMSE}}$ is generated by slicing x_p^{MMSE} to the nearest constellation point, i.e., $\hat{x}_p^{\text{MMSE}} = Q\left(x_p^{\text{MMSE}}\right)$. Thereafter, a detection search space is defined by expanding each scalar MMSE symbol with a number of neighbors. Specifically, for the ith symbol of the N-length MMSE vector $(\hat{x}_{p(i)}^{\text{MMSE}})$, a set of $(\Omega_i - 1)$ locally nearest sibling symbols is found:

$$x_{p(i)}^{\text{NB}} = [x_{p(i)}^1, \cdots, x_{p(i)}^{\omega}, \cdots, x_{p(i)}^{(\Omega_i-1)}], \tag{6.13}$$

with their distances to $\hat{x}_{p(i)}^{\text{MMSE}}$ sorted in ascending order, as

$$|x_{p(i)}^1 - \hat{x}_{p(i)}^{\text{MMSE}}|^2 \leq \cdots \leq |x_{p(i)}^{\omega} - \hat{x}_{p(i)}^{\text{MMSE}}|^2 \leq \cdots \tag{6.14}$$

Figure 6.5a and b illustrate the initial detection and search space delimitation for a case of 2×2 MIMO and 16-QAM modulation.

Once the search space is delimited, detection search paths are defined by generating a list $\mathcal{S}$ of candidate vectors using symbols drawn from the search space. Two methods exist for Candidate vector generation (CVG). First, occurrence of only

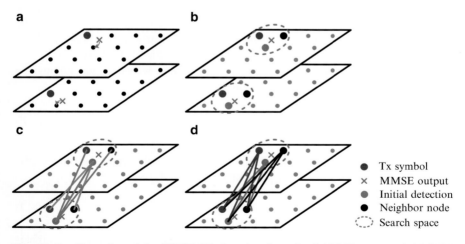

Fig. 6.5 An illustration of the MMSE-NP detection for a 2×2 MIMO setup and 16-QAM modulation. (**a**) Initial detection by slicing MMSE symbols to nearest constellation points. (**b**) Parallel node extension to include nearest neighbors into a search space. (**c**) Single-error candidate vector generation (SE-CVG). (**d**) Full-error candidate vector generation (FE-CVG)

one error is assumed during the initial detection. Accordingly, candidate vectors are generated by replacing only one symbol in $\hat{x}_p^{\mathrm{MMSE}}$ at a time, while keeping others unchanged, i.e., for the expanded $x_{p(i)}^{\mathrm{NB}}$, $(\Omega_i - 1)$ candidate vectors are generated as

$$
\begin{aligned}
s_i^1 &= [\hat{x}_{p(1)}^{\mathrm{MMSE}}, \cdots, x_{p(i)}^1, \cdots, \hat{x}_{p(N)}^{\mathrm{MMSE}}], \\
s_i^2 &= [\hat{x}_{p(1)}^{\mathrm{MMSE}}, \cdots, x_{p(i)}^2, \cdots, \hat{x}_{p(N)}^{\mathrm{MMSE}}], \\
&\vdots \\
s_i^{(\Omega_i-1)} &= [\hat{x}_{p(1)}^{\mathrm{MMSE}}, \cdots, x_{p(i)}^{(\Omega_i-1)}, \cdots, \hat{x}_{p(N)}^{\mathrm{MMSE}}].
\end{aligned}
\tag{6.15}
$$

After (6.15) being applied to all $x_{p(i)}^{\mathrm{NB}}$ ($i \in [1, N]$), $L = \sum_{i=1}^N \Omega_i$ candidate vectors are obtained in the list $\mathcal{S}$ including the initial MMSE result $\hat{x}_p^{\mathrm{MMSE}}$. In low-dimensional MIMO systems, such as 2×2, single-error dominates error events in the MMSE detection. However, for 4×4 or larger MIMO configurations, considering only one error in the initial detection is far from sufficient to cover most of the error events due to the increased degree of spatial selectivity. Hence, the second method considers a full-error scenario, i.e., assuming all symbols in x_p are erroneously detected. In consequence, all combinations of expended symbols $x_{p(i)}^{\mathrm{NB}}$ in $\mathcal{S}$ have to be included, resulting in totally $L = \prod_{i=1}^N \Omega_i$ candidate vectors to be searched. Figure 6.5c and d illustrate the Single-Error (SE-CVG) and Full-Error (FE-CVG) candidate vector generation schemes. Performance of symbol detection, in terms of Frame error rate (FER), using these two methods are compared in both 2×2 and 4×4 MIMO systems with 64-QAM modulation, see Fig. 6.6. As expected, performance of the SE-CVG scheme approaches to its full-error counterpart in the 2×2 MIMO system, whereas a large performance degradation is observed in the 4×4 case. In comparison, the FE-CVG scheme substantially improves the detection performance, i.e., by ~ 2 dB at FER $= 10^{-2}$ (see Fig. 6.6). Hence, this study adopts the FE-CVG scheme.

The final detection result is generated by searching within $\mathcal{S}$ and finding the vector with the smallest squared Euclidean distance (ED), i.e.,

$$
\hat{x}_p = \arg \min_{x_p \in \mathcal{S}} \left\| \tilde{y} - \underline{R} x_p \right\|_2^2.
\tag{6.16}
$$

The recovered transmitted vector $\hat{x}$ with its original symbol sequence is obtained by reordering the rows of $\hat{x}_p$ with the permutation matrix P, i.e.,

$$
\hat{x} = P \hat{x}_p.
\tag{6.17}
$$

Compared to conventional tree-search based algorithms, MMSE-NP eliminates scalar and data dependent operations, as symbol expansions, candidate generations, and evaluations are carried out in parallel on all spatial layers. As a result, it provides extensive DLP for efficient implementations on vector-based architectures.

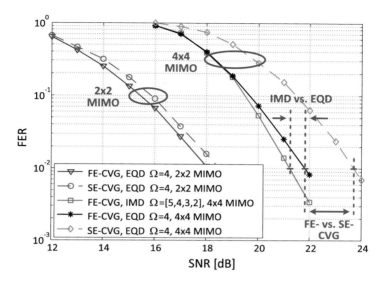

Fig. 6.6 Performance comparisons of different Candidate vector generation (CVG) and symbol expansion schemes

Imbalanced Node Perturbation

The perturbation parameter Ω_i in (6.13) needs to be adjusted to achieve a good performance-complexity trade-off. Basically there are two strategies to determine Ω_i. The first approach, referred to as *Equally distributed (EQD)* expansion, is to consider the same number of neighbors around each scalar symbol $\hat{x}_{p(i)}^{\mathrm{MMSE}}$, i.e., $\Omega_i = \Omega$. However, EQD expansion may not be cost effective from a complexity point of view, as channel properties of each antenna port is not utilized when determining the search space. Consequently, search paths in S may be over-selected, which increases computational complexity in (6.16) without any improvement in performance. Therefore, an *IMbalanced distributed (IMD)* expansion scheme is introduced to treat symbol expansion in $\hat{x}_{p(i)}^{\mathrm{MMSE}}$ differently and assign Ω_i depending on the channel condition. The idea is to include more neighbors for symbols located in spatial layers with lower post-detection SNRs (η in (6.10)), i.e., $\Omega_i > \Omega_j$ if $\eta_i < \eta_j$. These two expansion schemes are illustrated in Fig. 6.7 using the previous 2×2 MIMO setup.

Recall that $\hat{\boldsymbol{H}}$ is column-wise permuted during MMSE-SQRD with their corresponding η_i sorted in ascending order, i.e., $\boldsymbol{\eta} = [\eta_{\min}, \cdots, \eta_{\max}]$, the assignments of Ω_i are simplified by arranging the vector $\boldsymbol{\Omega}$ in descending order, namely $\boldsymbol{\Omega} = [\Omega_{\max}, \cdots, \Omega_{\min}]$.

Thanks to the use of channel properties, the IMD expansion scheme provides better performance than EQD while using fewer number of candidate vectors. As demonstrated in Fig. 6.6, detection using IMD with an expansion vector $\boldsymbol{\Omega} = [5, 4, 3, 2]$ is 0.6 dB better than the case of EQD with $\Omega = 4$ at FER $= 10^{-2}$,

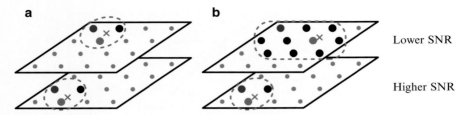

Fig. 6.7 Symbol expansion schemes, (**a**) Equally distributed (EQD) and (**b**) IMbalanced distributed (IMD)

even though the latter one uses two times more candidate vectors. Hence, the IMD expansion scheme is employed in this study. The exact $\mathbf{\Omega}$ assignment is a design parameter, which should be fine tuned at run-time.

Successive Partial Node Expansion (SPE)

The average error rate in a MIMO system is generally dominated by the spatial stream that suffers from the worst channel condition. Hence, node expansion for symbol with the smallest η value, i.e., $\hat{x}_{p(1)}^{\mathrm{MMSE}}$, needs to be handled with special care. According to the IMD expansion scheme, $\hat{x}_{p(1)}^{\mathrm{MMSE}}$ needs to be expanded with more neighbors to mitigate the high error probability. This strategy results in larger search space, which considerably increases the total number of candidate vectors, incurring huge computational complexity for the minimum-search process in (6.16).

To tackle this issue, a Successive partial node expansion (SPE) scheme is developed. It reduces the search space for $\hat{x}_{p(1)}^{\mathrm{MMSE}}$ without sacrificing detection performance. The basic idea is to utilize the property of the upper triangular matrix $\mathbf{R}$ and the fact that the symbol with $\eta_{\min}$ has been moved to the first layer after MMSE-SQRD. With $\underline{r}_{j,1}$ ($j = [2, \cdots, N]$) being zeros, the detection of $x_{p(1)}$ is solely dependent on $\tilde{y}_1$. Thereby, an optimal expansion of $x_{p(1)}$ can be obtained by simply solving a linear equation, given that other symbols have been expanded prior to $x_{p(1)}$. More specifically, SPE starts by expanding "stronger" symbols (i.e., $[\hat{x}_{p(N)}^{\mathrm{MMSE}}, \cdots, \hat{x}_{p(2)}^{\mathrm{MMSE}}]$) and then generates partial candidate vectors $\boldsymbol{x}_p^{[1]}$ of size $(N-1)$. Here, $\boldsymbol{x}_p^{[1]} = [x_{p(2)}, \cdots, x_{p(N)}]^T$ denotes the sub-vector of $\boldsymbol{x}_p$ with the 1st symbol $x_{p(1)}$ being omitted. Thereafter, $x_{p(1)}$ is obtained by substituting $\boldsymbol{x}_p^{[1]}$ into the 1st row of the system model (6.8):

$$\tilde{y}_1 = \sum_{j=1}^{N} \underline{r}_{1,j} x_{p(j)} = \underline{r}_{1,1} x_{p(1)} + \sum_{j=2}^{N} \underline{r}_{1,j} x_{p(j)} = \underline{r}_{1,1} x_{p(1)} + \underline{\boldsymbol{r}}_1^{[1]} \boldsymbol{x}_p^{[1]}$$

$$x_{p(1)} = \mathcal{Q}\left(\left(\tilde{y}_1 - \underline{\boldsymbol{r}}_1^{[1]} \boldsymbol{x}_p^{[1]}\right) \Big/ \underline{r}_{1,1}\right), \tag{6.18}$$

where $\underline{r}_1^{[1]} = [\underline{r}_{1,2}, \cdots, \underline{r}_{1,N}]$. For all L possible $x_p^{[1]}$ candidates, L number of $x_{p(1)}$ are found. This way, the search space for $\hat{x}_{p(1)}^{\mathrm{MMSE}}$ is reduced to only include symbols that, in conjunction with $x_p^{[1]}$, generate the most likely search paths, i.e., symbols that result in the smallest ED for the given $x_p^{[1]}$ vectors. Thus, the SPE scheme dramatically reduces the number of candidate vectors and thus the complexity of (6.16), while providing an equivalent performance as if all possible $x_{p(1)}$ symbols were included in the candidate vectors.

Due to the successive expansion of symbol $\hat{x}_{p(1)}^{\mathrm{MMSE}}$, the SPE scheme partially breaks the structure of the N-length vector $\hat{x}_p^{\mathrm{MMSE}}$. However, this adverse effect is substantially outweighed by the reduction of costly ED calculations and the efficiency of the optimal $\hat{x}_{p(1)}^{\mathrm{MMSE}}$ expansion.

Summary and Discussion

Figure 6.8 summarizes the computation procedure of the MMSE-NP algorithm. It contains four main processing stages: initial linear MMSE detection, IMD symbol expansion, ED calculation, and final detection. In Fig. 6.8, shaded boxes depict vector operations and layered boxes indicate parallel processing. $\circlearrowright$ (*l*) represents loops with count *l*, while V and M denote vector and matrix operations, respectively. As shown, most of the boxes are shaded in the figure, indicating a highly vectorized algorithm. Besides, the overall dataflow is regular, even though one loop structure is found in an inner block, i.e., symbol expansions of $[\hat{x}_{p(2)}^{\mathrm{MMSE}}, \cdots, \hat{x}_{p(N)}^{\mathrm{MMSE}}]$.

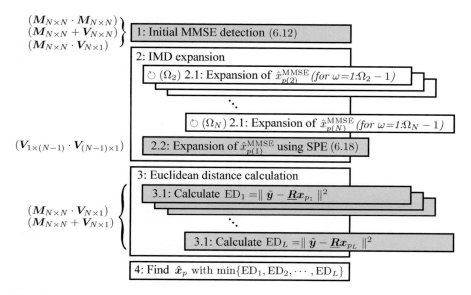

Fig. 6.8 Computation procedure of the MMSE-NP algorithm

6.3 Algorithm Evaluation and Operation Analysis

This section evaluates the adopted algorithms in terms of processing performance and computational complexity. As a metric for measuring performance, FER is used to show the effectiveness of the algorithms. Regarding computational complexity, the number of arithmetic operations is analyzed for data processing within one LTE-A time slot. Hardware friendliness is evaluated by analyzing the DLP and operation sharing of the three algorithms. Besides the operation analysis, task planning is conducted according to the timing specification of LTE-A.

6.3.1 Simulation Environment

Based on the structure of the MIMO-OFDM transceiver (Fig. 6.1), a simplified system setup is implemented in MATLAB with a special focus on the LTE-A system. Figure 6.9 shows the block diagram of the employed simulation environment. In the current setup, data are transmitted through a baseband spatially uncorrelated MIMO channel, where the maximum excess delay is smaller than the length of CP. Accordingly, domain (time-frequency) transformations and CP insertion/removing blocks are omitted. In addition, no pre-coding is implemented at the transmitter and MIMO systems are assumed to operate in a spatial-multiplexing mode. Moreover, the front-end block is omitted under assumptions of perfect front-end processing at the receiver, e.g., perfect synchronization and IQ-imbalance compensation.

Despite the system-level simplifications, the simulation setup is flexible as each block in Fig. 6.9 can be configured with various parameters. Based on the error correcting code specified in [2], a parallel concatenated turbo code [6] is adopted at the transmitter. Input parameters for this block are the coding rate and the interleaver block size. The generator polynomials are $g_0(D) = 1 + D^2 + D^3$ and $g_1(D) = 1 + D + D^3$. The modulation (mapping) block supports constellation sizes from BPSK to 64-QAM. Before inserting data and pilot tones into the LTE-A time-frequency grid, the layer mapping block maps encoded and modulated source data onto multiple antennas. Currently supported antenna sizes are 2×2 and 4×4.

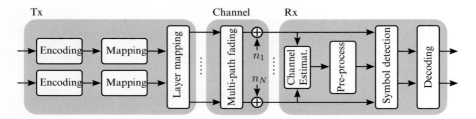

Fig. 6.9 Block diagram of the employed simulation environment

Table 6.1 Parameters for performance simulations in a LTE-A downlink

Block	Parameter	Value
Encode	Coding rate	1/2
	Interleaver block size	5376
Mapping	Constellation size	64-QAM
Channel	Antenna size	4×4
	Bandwidth	20 MHz
	Multi-path fading propagation	3GPP EVA
	Maximum Doppler frequency	70 Hz
	Time-variant/invariant	Quasi-static
Decode	Iteration number	6

The size of the time-frequency grid is determined by the allocated bandwidth. All bandwidth configurations specified in LTE-A are included, varying from 1.4 to 20 MHz. The propagation channel is modeled as a frequency-selective fading channel, in which its multi-path delay profile complies with the ones defined in the 3GPP specification [1]. Three channel models are supported, Extended pedestrian A (EPA), Extended vehicular A (EVA), and Extended typical urban (ETU). Besides, the maximum Doppler frequency is used for channel generations. Moreover, both time-invariant (i.e., quasi-static) and variant channel modeling are implemented. The former one assumes that channel coefficients remain unchanged within one LTE-A time slot, whereas the latter one emulates the scenario of constantly changing channels. The decoder at the receiver adopts the Bahl–Cocke–Jelinek–Raviv (BCJR) [3] algorithm with a configurable iteration number.

Detailed parameters used in the following performance simulations are summarized in Table 6.1. For each simulation, N_s LTE-A subframes (14 OFDM symbols) are transmitted, where N_s is dynamically adjusted to take account of different FERs with respect to SNR values. With a target of FER $= 10^{-2}$ that is a commonly used design criterion, N_s varies from 500 to 6000.

6.3.2 Performance Evaluation

Using the presented simulation environment, performance of the adopted MIMO processing algorithms are evaluated. It starts with analyzing the frequency correlation window (N_{SW}) in the R.MMSE-SW estimator and the node perturbation parameter (Ω) in the MMSE-NP detector. Off-line decisions on exact N_{SW} and Ω values are not required, as they can be fine-tuned at run-time thanks to the hardware flexibilities (Sect. 6.4). Therefore, the focus of the following analysis is to compare these algorithms to other alternatives. One example is to see whether the MMSE-NP approach bridges the algorithm–architecture gap between linear and tree-search based detectors. Additionally, the analysis also serves to make better design trade-offs, for example, by studying both positive- and side-effects of parameter variations.

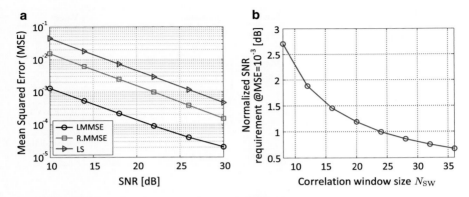

Fig. 6.10 Evaluation of channel estimator, (**a**) R.MMSE in comparison to LMMSE and LS algorithms, (**b**) comparison of different correlation window size N_{SW} in R.MMSE-SW

Channel Estimation

To minimize performance impacts from other processing tasks, the R.MMSE algorithm is firstly compared with LMMSE and LS estimator without involving channel matrix pre-processing and symbol detection. Figure 6.10a shows the performance comparison of the three estimation algorithms in terms of Mean squared error (MSE). It demonstrates that R.MMSE achieves a better estimation result than the LS approach, e.g., by 4.7 dB under this simulation setup, thanks to the reduction of noise enhancements and the use of second-order channel statistics. Compared to the LMMSE estimator, some performance degradation is observed, mainly due to the use of the underestimated function $\mathcal{W}$ (6.4). However, when considering its performance robustness and a huge complexity gain to the LMMSE method (i.e., more than two orders of magnitude as shown further in Sect. 6.3.3), the R.MMSE approach is more attractive to implement in practice, especially for resource- and energy-limited devices.

Using the same simulation setup, Fig. 6.10b compares R.MMSE-SW and R.MMSE with respect to different N_{SW} values. Numbers at the vertical axis denotes the minimum SNR required to reach the level of MSE $= 10^{-3}$, normalized to the full-window case R.MMSE. Large values of N_{SW}, as expected, lead to small performance degradation in comparison to the R.MMSE case, but result in high computational complexity.

Based on these analysis, the adopted R.MMSE-SW estimator is further evaluated on a system-level, namely by including succeeding channel matrix pre-processing and symbol detection and measuring the output FER. Meanwhile, required coefficient ROM sizes are calculated to illustrate (to some extent) hardware costs with respect to different N_{SW} values. A more detailed operation analysis is further presented in Sect. 6.3.3. To avoid any influence between R.MMSE-SW and MMSE-NP detector, the conventional near-ML FSD algorithm is used. Figure 6.11 shows the achieved FERs versus (ROM) sizes for various N_{SW} values. Both coordinates

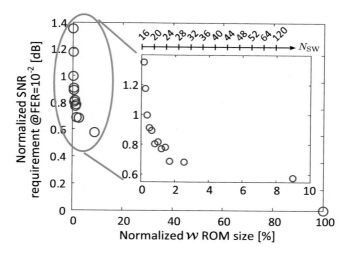

Fig. 6.11 Performance versus coefficient ROM size for different N_{SW} values in the R.MMSE-SW estimator. Metrics are normalized to that of the full-window case—R.MMSE, which has 100 % ROM size and zero required SNR at FER $= 10^{-2}$

are normalized to the reference case R.MMSE. It clearly shows that small values of N_{SW} reduce ROM size substantially while retaining a good system performance. As an example, with $N_{SW} = 24$, the required ROM size is reduced by 99.64 % compared to that of the R.MMSE, at the cost of less than 1 dB FER degradation. These differences in FER diminish with increasing N_{SW} values. To conclude, N_{SW} should be fine-tuned at run-time to achieve on-demand performance-complexity trade-offs.

Symbol Detection

In this section, the node perturbation parameter $\mathbf{\Omega}$ in the MMSE-NP detector is evaluated in comparison to conventional linear and tree-search based detection algorithms, e.g., "linear MMSE" and "K-Best and FSD", respectively. To minimize performance impacts, these detectors are evaluated without performing other processing tasks. In other words, it is assumed that channel knowledge is perfectly estimated at the receiver and that channel matrices are properly processed by performing inversion (with the MMSE criterion) for the linear MMSE detector and QRD for the K-Best and FSD cases.

Figure 6.12 shows simulated FERs of different detection algorithms. For the MMSE-NP detector, the notation $\mathbf{\Omega} = [F, \cdots]$ represents the employed SPE scheme (Sect. 6.2.3). Thanks to the developed techniques in MMSE-NP, i.e., node perturbation, IMD, and SPE, the performance of the linear MMSE detector is enhanced substantially. More importantly, an FER performance close to that of the

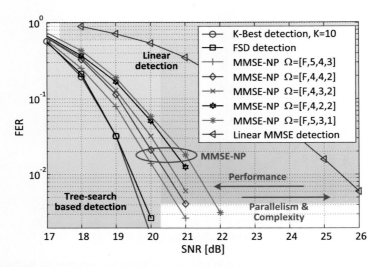

Fig. 6.12 Comparison between linear, tree-search based, and the employed detection algorithms

K-Best detector and FSD is achieved. For $\boldsymbol{\Omega} = [F, 5, 4, 3]$, performance degradation to both K-Best decoder (with $K = 10$) and FSD is less than 1 dB at FER $= 10^{-2}$. Better performance is obtained by including more candidate vectors in the symbol expansion at the expense of implementation complexity. This is similar to the tree-search based detectors with different number of branch traversals, e.g., K-Best algorithm with different K values. Figure 6.12 also compares the FER of different $\boldsymbol{\Omega}$ assignments. Comparing the cases $\boldsymbol{\Omega} = [F, 5, 4, 3]$ and $\boldsymbol{\Omega} = [F, 5, 3, 1]$, the former one is 1 dB better than the latter case, but with four times more candidate vectors involved in detection, thus demands more computational power.

Using the number of visited nodes as a first-order complexity analysis, Table 6.2 summaries the performance metrics for the four algorithms. Based on the node perturbation scheme, the node expansion number of the MMSE-NP detection is formulated as

$$N^{\text{MMSE-NP}} = \sum_{i=1}^{N} \Omega_i N_{i+1} = \sum_{i=1}^{N} \Omega_i \left(\prod_{j=i+1}^{N} \Omega_j \right), \tag{6.19}$$

where N denotes the number of antennas, N_i is the number of nodes at the ith spatial stream, and $N_1 = \Omega_1 = 1$ when using the SPE scheme. The total number of visited nodes in the K-Best algorithm [33] is calculated as

$$N^{\text{K-Best}} = M \sum_{i=1}^{N} N_F^{i+1}, \tag{6.20}$$

Table 6.2 Comparison of visited nodes and required SNR at FER $= 10^{-2}$

	Parameter	N_{visited}		SNR [dB] @FER $= 10^{-2}$	
K-best	$K = 10$	1984	$-$ (ref.)	19.39	0
FSD	$P = 1$	256	7.75×	19.47	+0.08
MMSE-NP	$\Omega = [F, 5, 4, 3]$	135	14.70×	20.20	+0.81
	$\Omega = [F, 5, 3, 1]$	34	58.35×	21.34	+1.95
MMSE	N/A	N/A	N/A	25.48	+6.09

where $N_F^i = \min(K, MN_F^{i+1})$ denotes the number of parent nodes at the ith layer with M being the constellation size. For the FSD [4], the number of visited nodes is

$$N^{\text{FSD}} = \sum_{i=1}^{N} \prod_{j=i}^{N} p_i. \tag{6.21}$$

It shows in Table 6.2 that the number of nodes visited in the MMSE-NP algorithm with $\Omega = [F, 5, 4, 3]$ is 15 and 1.9 times fewer than that of the K-Best detector and FSD, respectively, which demonstrates the cost effectiveness of the MMSE-NP. In summary, the presented MMSE-NP algorithm bridges the algorithm–architecture gap between linear and tree-structured detection schemes. In addition, with imbalanced Ω assignments, the algorithm is highly scalable, since the symbol detection of each spatial stream can be tuned dynamically to adapt to instantaneous channel condition or currently available computational resources.

MIMO Signal Processing

After the analysis of individual algorithms, the MIMO processing tasks are evaluated together by using different combinations of algorithms. This is aimed to compare the employed processing scheme, "R.MMSE-SW+MMSE-NP," with other approaches with respect to performance and computational complexity. For the following analysis, parameters of $N_{\text{SW}} = 24$ and $\Omega = [F, 4, 3, 2]$ are used for R.MMSE-SW and MMSE-NP, respectively. Figure 6.13 gives a full picture of the performance-complexity trade-offs for different algorithm sets. For better illustration, they are grouped into three clusters based on the involved channel estimation method. In Fig. 6.13, numbers at the vertical axis denotes the minimum SNR required to achieve the target 10^{-2} FER, while the computational complexity measured in the number of operations required in one LTE-A time slot is shown along the horizontal axis. In addition, both coordinates are normalized to a reference case, "LMMSE+FSD", which provides the best performance among these algorithms. According to this setup, algorithms whose coordinates are closed to the bottom-left corner are desired.

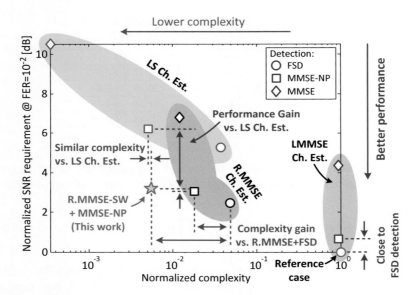

Fig. 6.13 Analysis of processing performance and computational complexity. Metrics are normalized to that of the reference case "LMMSE+FSD" which has unit computational complexity and zero required SNR at FER $= 10^{-2}$

It shows in Fig. 6.13 that the adopted scheme "R.MMSE-SW+MMSE-NP" achieves a good design trade-off between performance and complexity. For instance, it provides more than 7 dB performance gain to the "LS+MMSE" method (upper-left corner) and achieves more than two orders of magnitude complexity reduction to the "LMMSE+FSD" case (bottom-right corner). It should be re-emphasized that N_{SW} and Ω are tunable parameters and should be optimized at run-time. In the following section, DLP, operation sharing, and computational complexity of the three tasks are analyzed in detail.

6.3.3 Operation and Complexity Analysis

With the presented algorithms, primitive operations required by the R.MMSE-SW estimator, MMSE-SQRD pre-processor, and MMSE-NP detector are characterized. Table 6.3 summarizes required vector and scalar operations and their proportion in each task. Two meaningful properties can be observed. First, most of the operations are at vector-level thanks to the development of algorithm vectorization. Specifically, vector operations occupy more than 98 % of the total workload in all three tasks, indicating high DLP. This is an important design criterion for attaining efficient implementations with respect to processing throughput and energy consumption. Second, most of the primitive operations are shared among

Table 6.3 Algorithm profiling for primitive vector (V) and scalar (s) operations in the adopted MIMO signal processing

	Primitive operation	Operation dimension and proportion in each task						Total proportion [%]
		R.MMSE-SW		MMSE-SQRD		MMSE-NP		
Vector	$A \odot B^{a}$	–	–	$V_{(N \times 1)}$	35 %	–	–	4.30
	$A \cdot B$	$V_{(N_{sw} \times 1)}$	91 %	$V_{(N \times 1)}$	35 %	$V_{(N \times 1)}$	84 %	80.98
	$A \pm B$	–	–	$V_{(N \times 1)}$	15 %	$V_{(N \times 1)}$	15 %	13.36
Scalar	$x_a \cdot x_b$	$s(x_a \cdot x_b)$	9 %	–	–	–	–	0.29
	Sorting	–	–	$s(x_i)$	5 %	$s(x_i)$	~ 0 %	0.07
	$1/\sqrt{x}$	–	–	$s(x)$	10 %	–	–	0.14
	Pert.^{b}	–	–	–	–	$s(\Omega_i)$	1 %	0.86

^{a} $\odot$ Element-wise vector multiplication
^{b} Node perturbation in symbol detection

these algorithms, implying the potential of extensive hardware reuse. Moreover, when considering a vector dot product as an element-wise vector multiplication followed by a vector addition, all vector operations are common to all three tasks.

Based on the operation profiling, computational complexity of the algorithms is analyzed. To simplify the analysis, same precision is assumed for all calculations and a W-bit complex-valued addition is used as a baseline operation. This way, a W-bit complex-valued multiplication has the complexity of W; a W-bit real-valued division and square root operation has a complexity of KW with K being a scaling factor, e.g., iteration number in Newton–Raphson method [16], and is set to 2 in this study.

Given these assumptions, complexity of "R.MMSE-SW+MMSE-NP" is compared with other algorithms in Fig. 6.13. As for the three cases inside the LMMSE group, the complexity of the LMMSE algorithm is so dominating that it almost conceals any difference between different channel matrix pre-processing and symbol detection algorithms. By comparison, the adopted R.MMSE-SW algorithm shows a similar level of complexity to that of the LS estimator, while providing much better processing performance. In terms of symbol detection, the employed MMSE-NP algorithm demonstrates 2.7 times complexity reduction compared to FSD. The combination of R.MMSE-SW and MMSE-NP is 8.6 times less complex than the "R.MMSE+FSD" case, at the price of less than 1 dB performance degradation. In summary, the employed processing scheme provides a good performance-complexity trade-off and is hardware friendly to vector-based architectures.

6.3.4 Processing Flow and Timing Analysis

In this chapter, dataflow and timing analysis is presented according to the LTE-A specification. Table 6.4 summaries resource allocations in the 20 MHz LTE-A. Within one time slot (0.5 mS), 7 OFDM symbols are transmitted, including 46.88 %

Table 6.4 Resource allocations in one LTE-A time slot

Time slot (t_{slot})	0.5 ms		
Bandwidth	20 MHz		
Sampling frequency	30.72 MHz		
Number of subcarriers/symbol	2048		
Total number of subcarriers	14,336	466.67 μs	93.33 %
Total length of CP	1024	33.33 μs	6.67 %
Data-carrying subcarriers	7200	234.38 μs	46.88 %
Pilot tones	1200	39.06 μs	7.81 %
Guard band subcarriers	5936	193.23 μs	38.65 %

of data, 7.81 % of pilots, 38.65 % of guard-band subcarriers, and 6.67 % of cyclic prefix. Figure 6.14a shows the structure of a 4×4 MIMO LTE-A data frame and the flow of target baseband processing. As illustrated, the LS computation in (6.2) can be initiated as soon as the pilot data has been received, followed by the frequency-domain interpolation (6.4). However, channel matrix pre-processing and subsequent symbol detection cannot start until the second OFDM symbol is received due to pilot receptions for antenna ports 2 and 3 (Fig. 6.2). As a consequence of this sequential processing flow, one can see from Fig. 6.14a that processing gaps widely exist in between neighboring OFDM symbols as well as consecutive time slots. Thus, implementations using task-dedicated hardware will result in poor resource utilization. Moreover, the cyclic prefix and guard band interval between adjacent OFDM symbols further enlarge those processing gaps.

To attain efficient hardware usage, this study maps three tasks on one recon-figurable platform by utilizing the sequential nature of processing and non-data-carrying time intervals. Figure 6.14b illustrates a task-oriented processing flow, which performs one task on all subcarriers before switching to the subsequent one. This is different from a subcarrier-oriented scheme (i.e., handling one subcarrier at a time), which requires much more power to carry out frequent context switching. According to [30], dynamic configurations may take up to 40 % of the overall power consumption in reconfigurable platforms. Thereby, reducing the number of context switching is an efficient way to achieve power efficient implementation. Note that every processing iteration shown in Fig. 6.14 starts immediately the last pilot tone in OFDM symbol 1 is received. This is arranged to prevent processing gaps due to data awaiting by making sure that all required pilot tones have been buffered.

Baseband processing in LTE-A systems requires data buffering of several OFDM symbols [11] to, for example, cope with the orthogonal pilot pattern and processing latency of control channels [53]. Thus, additional buffers are not required in the adopted solution if all the three processing tasks can be handled within the specified time interval. In this study, processing is scheduled on a basis of one LTE-A time slot, see Fig. 6.14. Therefore, the computation time of each iteration (t_{iter}) is constrained by t_{slot}, such that $t_{iter} \leq t_{slot} = 0.5$ mS. This is used as a design constraint to guide hardware development.

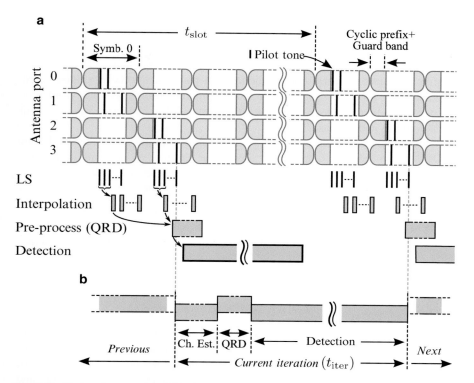

Fig. 6.14 Timing diagram of the MIMO signal processing, (**a**) an LTE-A frame structure and data dependency between processing tasks, (**b**) adopted task-oriented processing flow

6.4 Hardware Development

Using the reconfigurable cell array developed in Chaps. 4 and 5 as a baseline architecture, this section presents a number of hardware enhancements for attaining efficient implementation of MIMO signal processing. The focus here is on vector computing using heterogeneous Resource cells (RCs) and various memory access schemes. In addition, a technique for further improving processing throughput and hardware efficiency is elaborated.

Before presenting the architectural development, three main properties of MIMO signal processing are extracted from the aforementioned operation analysis. Correspondingly, hardware requirements are identified with respect to *computation*, *memory access*, and *data transfer*.

- **Massive vector operations**: in view of the massive vector operations, i.e., more than 98 % of the total workload in Table 6.3, efficient vector computing and high bandwidth memory access are essential. Besides, it is beneficial to reduce the number of register/memory accesses and data transfers to keep processing

overhead low, since the control (regarded as a part of control overhead) required
for performing those operations may consume a large portion of time and
power [32].

- **Hybrid data-widths and formats**: the coexistence of scalar and vector operations requires a hybrid computational data path. Additionally, efficient communication mechanisms are expected to offload processing units from non-
computational operations, e.g., data alignments, during data transfers of various
data-widths and formats.
- **Multi-subcarrier processing**: as a scheduling technique for further exploiting
DLP (Chap. 3.3), multi-subcarrier processing requires various data access patterns to perform operations simultaneously at multiple subcarriers. Therefore,
flexible memory access schemes are required, e.g., concurrent access of vectors
from different channel matrices.

These requirements pose design challenges for hardware development and need
to be addressed during the architectural design to ensure implementation efficiency.

6.4.1 Architecture Overview

Built upon the baseline architecture (Fig. 6.15a), the baseband processor is composed of four heterogeneous tiles, which are partitioned into scalar- and vector-
processing domains to cope with hybrid data computing, see Fig. 6.15b. In the vector
domain, Tile-0 handles vector processing while Tile-1 provides data storage and
various forms of vector and matrix accesses. In the scalar domain, Tile-3 controls
other RCs during run-time and handles scalar and irregular operations with memory
supports from Tile-2. Data transfers between the two domains are bridged by
memory cells using the micro-block function (Sect. 5.3.2). This feature efficiently
supports hybrid data transfers without additional controls from processing cells. For
example, memory cells in Tile-1 provide vector data accesses to RCs in Tile-0 while
exchanging data in a scalar form with RCs in Tile-2.

Besides the heterogeneous resource deployments, communication to an external
host for both data transfers and off-line configurations are carried out using the
hierarchical network. Run-time configurations for all RCs are issued on a per-clock-
cycle basis, performed hierarchically within the cell array, and managed jointly by
a task manager (i.e., a processing cell in Tile-3) and local controllers distributed
in RCs. Specifically, the task manager tracks the overall processing flow, controls
context switching (e.g., changing from channel estimation to pre-processing), and
handles configuration updating (e.g., parameter updates for N_{SW} and Ω). Local
controllers are responsible for applying configurations onto processing data- and
memory access-paths to, for example, switch between operations listed in Table 6.3.

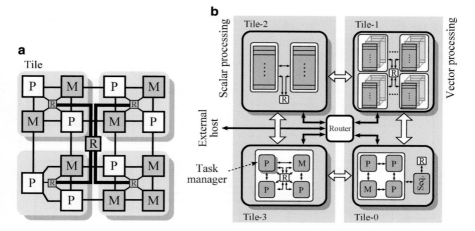

Fig. 6.15 (**a**) Baseline architecture of the reconfigurable cell array, an example of four tiles. (**b**) Block diagram of the employed heterogeneous baseband processor. Distributed controllers within Resource cells (RCs) are omitted in the figure for simplicity

6.4.2 Vector Dataflow Processor

Figure 6.16 shows the architecture of Tile-0, a vector dataflow processor, consisting of three processing cells (pre-, core-, and post-processing), one memory cell (register bank), and a sequencer. The three processing cells, shown on the upper half of Fig. 6.16, are deployed for vector computations. The register bank provides data accesses from both internal registers and other tiles through register-mapped IO ports. The sequencer controls operations of the other cells via a control bus, drawn in dashed lines in Fig. 6.16.

Atomic operations of Tile-0 are built upon N-length vectors which is the most common data type of the vector processing in Table 6.3. Vectors exceeding this length are processed by folding, i.e., they are decomposed into data segments suitable for atomic operations. Local data transfers within Tile-0 are carried out on two $N \times N$ matrix and one $N \times 1$ vector bus, arranged both to suffice computational requirements and to improve processing efficiency. The two matrix buses are used to support data intensive operations such as data-tone MMSE interpolation in (6.4) and Euclidean norm ($\| \cdot \|$) of augmented channel matrix $\hat{\boldsymbol{H}}$ in (Algorithm 1 lines 3 and 12), both requiring two $N \times N$ matrix inputs. The vector bus, on the other hand, is used to accelerate three-input operations such as column vector update in (Algorithm 1 line 11), which would otherwise require twice of the clock cycles with additional operations for loading and storing intermediate results.

In the following sections, the adopted configurable Instruction set architecture (ISA) and two vector processing enhancements are presented.

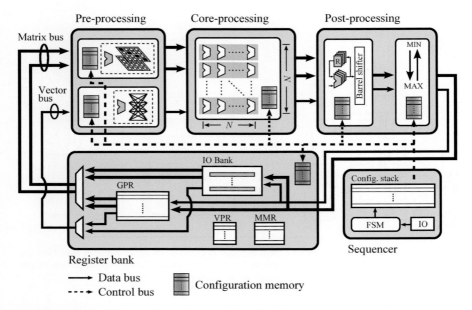

Fig. 6.16 Microarchitecture of the vector dataflow processor (Tile-0). A VLIW-style multi-stage computation chain consists of three processing cells: pre-, SIMD vector core-, and post-processing

Configurable Instruction Set Architecture

Conventionally, processors are implemented based on fixed ISAs, e.g., the generic and dataflow processors presented in Chaps. 4 and 5. Depending on target applications, the ISAs contain different specifications of, for example, instructions and addressing modes, and cannot be changed once they have been implemented. As a consequence, tasks outside the set originally intended may not benefit from the available computing capability, since underlying data- and control-path are hidden inside the fixed ISAs. Therefore, this design strategy often results in either limited flexibility (the case of application-specific) or poor performance (the case of general-purpose). In addition, it may require a deep study of target applications, which may not always be possible, concerning time-to-market, adoption of new algorithms, etc.

Some processors adopt configurable ISAs, which can be customized for different use. Configurations can be performed either during the chip synthesis [8] or at run-time using a similar approach to FPGAs [29, 47]. In view of the run-time configurability, the latter approach is desired. However, it is commonly implemented using a centralized control scheme, which incurs high control overhead with regard to configuration time, hardware complexity, and storage requirements. As an example, Fig. 6.17 shows the configurable ISA of a FlexCore [47]. The processor exposes its entire data- and control-path to the user via a long instruction word, 91 bits in this example. A controller controls dataflow and operations of each

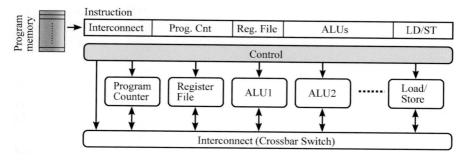

Fig. 6.17 Illustration of a fine-grained centralized control scheme in a configurable ISA, an example of FlexCore [47]

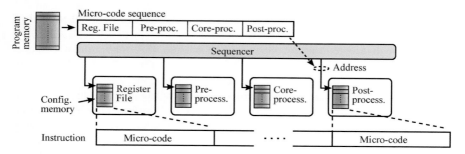

Fig. 6.18 Illustration of the employed distributed micro-code execution scheme in the vector dataflow processor

hardware unit based on instructions fetched from a program memory. It is inefficient that configurations of all RCs are centralized in one instruction, since any change among those configurations requires loading of a whole new instruction, resulting in unnecessary program storage and memory access for unchanged parts. Many code size reduction schemes exist which have reported a maximum compression ratio of about 70 % on a Very long instruction word (VLIW) processor [51]. However, this reduction comes at an area cost of up to 30 % for run-time instruction decompression.

To tackle the aforementioned overhead issue, two control techniques are employed in the adopted run-time configurable ISA.

Distributed Micro-Code Execution Figure 6.18 illustrates a distributed control scheme employed in this study to reduce the overhead of the long instruction word. The idea is to divide an instruction into a number of smaller ones, termed as micro-codes, each getting dispatched to an RC. For storing the micro-codes, each RC is deployed with a configuration memory, which can be accessed individually without affecting others. The size of these memories can be kept small, since the number of operations required from each RC in an application is often limited. In addition, new micro-codes can be prepared and loaded to memories while current instructions are being executed. This further reduces the storage requirement of the configuration memories. Using this distributed control scheme, fetching an instruction only involves address managements of the configuration memories. The required list of memory addresses is referred to as a micro-code sequence. Compared to the size of a micro-code, a memory address has much smaller wordlength, thus reducing the control overhead.

To demonstrate the gain of this scheme, a numerical example is given as follows. Considering a case where each configuration memory in Fig. 6.18 is of size 32bit × 16. The corresponding wordlength required for fetching an instruction when using the conventional approach is $32 \text{ bits} \times 4 = 128$ bits. With the presented scheme, a micro-code sequence requires only $4 \text{ bits} \times 4 = 16$ bits, reducing the wordlength of the program memory by eight times. In the case of storing $D = 256$ instructions, the reduced wordlength leads to a further memory reduction of 5.3 times, since only $(16 \text{ bits} \times D + (32 \text{ bits} \times 16) \times 4)$ bits are required instead of $(128 \text{ bits} \times D)$ bits.

Using the distributed micro-code execution scheme, ISA configuration contains two steps. First, micro-codes of individual RCs need to be defined and loaded to the distributed configuration memories. Second, micro-code sequences need to be specified and stored in the program memory. The complete micro-code set for each processing cell in Tile-0 is presented in Appendix B. Worth mentioning is that no branching instructions (except loops which are treated differently, see the following section) are implemented in the vector dataflow processor, since the processor is intended to be used for data-centric stream processing that often has exposed data dependencies and deterministic processing structure. Computation tasks mapped onto the processor are performed by invoking a series of kernel functions, such as matrix multiplication and QR decomposition. Switching between these functions are conducted by the generic processing cell in Tile-3. This way, program branching and the execution of conditional operations are mimicked. The assisted instruction branching is illustrated in Fig. 6.19.

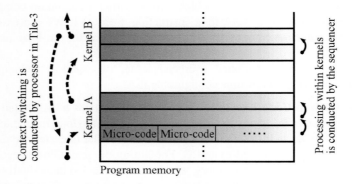

Fig. 6.19 Assisted instruction branching in the vector dataflow processor

Multi-Level Zero-Delay Inner Loop Control To further reduce control overhead during loop operations, a multi-level inner loop control scheme is adopted. Section 4.3.1 presents an inner loop controller designed to conduct loop operations with a zero execution latency. However, it supports only one loop level, far from sufficient for performing baseband processing in MIMO-OFDM systems. Multi-level loops are widely used to process multiple subcarriers per OFDM symbol, multiple spatial layers per subcarrier, and multiple iterations per spatial layer. Hence, the zero-delay one-level loop control scheme is extended to efficiently process loop operations without limit on the loop hierarchy.

To achieve the multi-level loop control, a stack-based architecture is employed, illustrated in Fig. 6.20a. As shown, the loop controller contains a configuration stack used to store the address of the first instruction in a loop (link address) and the corresponding loop count. Compared to other memory structures, the stack has a simple control mechanism. It natively supports the execution order of multi-level loops. With the help of a Finite-state machine (FSM), link addresses of loops are pushed into the stack in a "last-in-first-out" manner during the execution of a program. Figure 6.20b shows a snapshot of the stack when the entire loop hierarchy of the enclosed code fragment is pushed into the stack. Each instruction contains a flag used to indicate end-of-loop, similar to the one used in the one-level loop controller (Sect. 4.3.1). Upon the completion of a loop iteration, the controller updates the program counter with the link address drawn from the top of the stack in order to jump back to the start of the loop. Meanwhile, the stored loop count decreases by 1. When the counter value reaches 0, a loop operation is completed and automatically popped out, the stack pointer decreases by 1, and the link address stored at the new stack top is fetched. When the bottom of the stack is reached, loop controller releases the control of the program counter and the subsequent instruction stored in the program memory is fetched for execution. As can be seen, this enhanced loop control scheme requires no loop management operations from the user. Therefore, it eases program writing, speeds up loop processing, and reduces program size and control overhead.

Based on the configurable ISA, the following sections focus on data path of Tile-0 and present architectural improvements for attaining efficient processing.

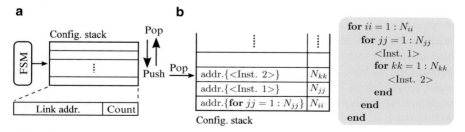

Fig. 6.20 (a) Multi-level inner loop control. (b) A snapshot of the configuration stack and an example of a code fragment

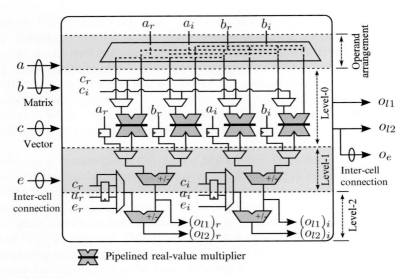

Fig. 6.21 Architectural diagram of a CMAC unit

Vector-Enhanced SIMD Core

In wireless baseband processing, Single instruction multiple data (SIMD) is commonly used as a baseline architecture to exploit inherent DLP. Similarly, a SIMD-based architecture is adopted in the core-processing cell, containing $N \times N$ homogeneous Complex-valued multiply-accumulate (CMAC) units (Fig. 6.16). The two-dimensional CMAC bank is deployed to handle parallel MIMO data streams and perform all vector operations in Table 6.3. Figure 6.21 shows a detailed architecture of the CMAC unit, containing four data inputs, three levels of arithmetic elements, and an input operand arrangement unit. With data inputs $\{a, b\}$ and $\{c\}$ coming from the matrix and vector data bus respectively, arithmetic elements in the first two levels are used to conduct complex-valued multiplication and addition. Two adders in level-2 sum up level-1 outputs with different data operands, such as a, c and e, depending on data path configurations.

Concerning the execution latency of vector operations, conventional SIMD architectures (e.g., [8, 32]) are inefficient, since they are designed to handle parallel independent scalar data operands and their function units between processing lanes cannot operate collaboratively during instruction execution. For example, the computation of Vector dot product (VDP), which takes more than 80 % of entire vector processing in Table 6.3, requires multiple clock cycles (depending on vector length), since each efficiently mapped VDP operation is performed in a folded fashion using at most one CMAC unit. This not only increases execution latency but also causes a large number of suspended computational resources. Although concurrent operations on multiple data sets may alleviate the latency issue to some extent, they require additional data buffers for storing intermediate results and a

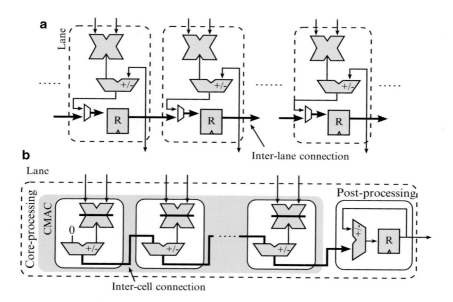

Fig. 6.22 (**a**) Conventional scalar-based SIMD architecture [31]. (**b**) Illustration of a processing lane in the vector-enhanced SIMD core

more sophisticated sequence control. Figure 6.22a shows a typical scalar-based SIMD architecture, where data transfers between processing lanes are only possible through internal registers.

In contrast, this study tackles the latency issue by adopting an effective low-complexity vectorization technique in the conventional SIMD architecture. This vector enhancement enables single-clock-cycle execution for all vector operations of length N. Specifically, each processing lane is expanded to have N CMAC units, each of which is equipped with an inter-cell connection (e-path in Fig. 6.21) to link up with neighboring CMACs during instruction execution. For example, the e input in Fig. 6.21 is connected to the level-2 output (O_e) of the previous CMAC unit. Using this simple connection, level-2 adders of CMACs in every processing lane can be concatenated to form an adder-tree capable of computing one N-length vector in every clock cycle, e.g., a VDP with an atomic operation of '$ab + e$'. Figure 6.22b illustrates the construction of the adder-tree using the vector-enhanced processing lane. For practical implementations, a balanced tree structure (not shown in Fig. 6.22b) is used to reduce the critical path of the SIMD core. Vectors exceeding the length N are processed by folding. In other words, they are decomposed into data segments suitable for atomic operations. The net results of this vector enhancement are significantly reduced execution latency and simplified sequence control.

Besides the efficient VDP computing, numerous vector operations are supported by the SIMD core, e.g., vector addition/subtraction '$a \pm b$' (6.16) and multiply-add '$a \pm bc$' (Algorithm 1 line 11). This is achieved by utilizing the flexible structure of

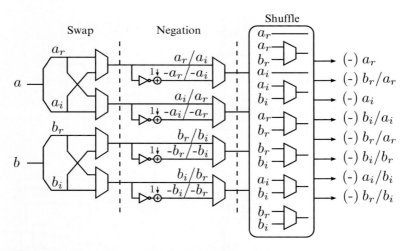

Fig. 6.23 Block diagram of the input operand arrangement unit

Table 6.5 Some commonly used operations and the corresponding data sequences generated by the input operand arrangement unit

Operation	Expression	Data sequence
Complex-MUL	$(a_r + ja_i) \cdot (b_r + jb_i)$	$a_r, b_r, a_i, b_i, a_r, b_i, a_i, b_r$
Complex-MUL with j	$(j(a_r + ja_i)) \cdot (j(b_r + jb_i))$	$-a_i, -b_i, a_r, b_r, -a_i, b_r, a_r, -b_i$
Complex-MUL with $-$j	$(-j(a_r + ja_i)) \cdot (-j(b_r + jb_i))$	$a_i, b_i, -a_r, -b_r, a_i, -b_r, -a_r, b_i$
Complex-squared norm	$(a_r + ja_i) \cdot (a_r + ja_i)^H$	$a_r, a_r, a_i, a_i, b_r, b_r, b_i, b_i$
	$(b_r + jb_i) \cdot (b_r + jb_i)^H$	
Real-MUL	$a_r b_r, a_i b_i$	a_r, b_r, a_i, b_i
Real-square	$a_r^2, a_i^2, b_r^2, b_i^2$	$a_r, a_r, a_i, a_i, b_r, b_r, b_i, b_i$

CMAC units, in which each level of the arithmetic elements can be used individually or operated with different combinations of data operands. To further extend the operation set, an operand arrangement unit is deployed at input of each CMAC unit, see Fig. 6.23. It contains three main function blocks, capable of swapping, negating, and shuffling the real and imaginary part of input operands **a** and **b**, respectively. Giving this flexibility, various data sequences are provided to the following CMAC unit for performing complex- and real-valued operations. Using a_r, a_i, b_r, and b_i to denote the real and imaginary part of operands **a** and **b** respectively, Table 6.5 lists some of the data sequences required by the commonly used operations.

VLIW-Style Multi-Stage Computing

Another important observation from the algorithm analysis (Sect. 6.3) is that most of the vector processing involve several tightly coupled operations, such as complex conjugate (Algorithm 1 line 10) and result sorting (6.16) performed, respectively, before and after vector computations. Mapping of such "long" processing solely on the SIMD core requires multiple atomic operations, causing not only increased execution time but also redundant register file accesses for intermediate result buffering. Moreover, execution of some operations, such as complex conjugate, only uses a small part of the function units, resulting in poor resource utilization. Hence, the SIMD core is extended by adopting a VLIW-style multi-stage computation chain to accomplish several consecutive data manipulations in one single instruction. Specifically, two distinct processing cells are arranged around the core-processor to pre- and post-process data respectively, see Fig. 6.16. Benefiting from this arrangement, more than 60 % of register accesses are avoided, as the combination of pre- and post-processing takes about two-thirds of the total vector computations. As an example, Table 6.6 summarizes operations required for implementing the MMSE-SQRD algorithm. A similar technique named operation chaining for reducing register accesses can be found in [32].

Implementation of the pre- and post-processing cells depends on the operation profile of target applications. In the case of MIMO signal processing, the pre-processing cell consists of two function units that, respectively, work with matrix and vector data. Data negation and absolute calculation are examples of the pre-processing operations, which can be applied individually to each part (real and imaginary) of complex-valued data operands. For matrix inputs, a matrix data mask function is adopted to ease the run-time generation of regular and frequently used data access patterns, e.g., construction of the identity matrix required by $\hat{\underline{H}}$ (6.6). Matrix data masks are stored in Matrix mask register (MMR), see Fig. 6.16. Each mask contains a boolean data map, used to indicate the "existence" of the matrix element at the corresponding position. The masking operation is realized by logically ANDing the matrix input with the data mask, real and imaginary part separately, illustrated in Fig. 6.24. Examples of some commonly used masks are identity-,

Table 6.6 An example of the multi-stage computing in MIMO channel matrix pre-processing (MMSE-SQRD, Algorithm 1)

Pre-processing		Pre-1: complex conjugate	
		Pre-2: vector shuffling & broadcast	
		Pre-3: matrix data masking	
Post-processing		Post-1: barrel shifting	
		Post-2: sorting	
Operation	Pre-processing	Core-processing	Post-processing
$\xi_i = \|\underline{q}_i\|_2^2$ & sort	Pre-3	VDP $(ab + e)$	Post-1, 2
$\underline{r}_{i,i} = \sqrt{\xi_i}$	–	VDP $(ab + e)$	Post-1
$\underline{q}_i = \underline{q}_i / \underline{r}_{i,i}$	Pre-2	bc	Post-1
$\underline{r}_{i,k} = \underline{q}_i^H \underline{q}_k$	Pre-1	VDP $(ab + e)$	Post-1
$\underline{q}_k = \underline{q}_k - \underline{r}_{i,k}\underline{q}_i$	Pre-2	$a - bc$	Post-1
$\underline{R}^{-1} = 1/\sigma_n \underline{Q}_b$	Pre-2	bc	Post-1

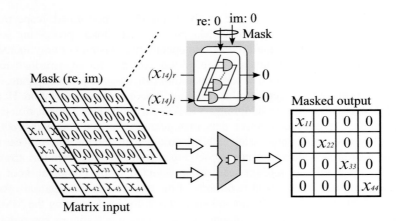

Fig. 6.24 Illustration of matrix masking operation, an example of the diagonal matrix construction

diagonal-, and upper triangular-matrix, and real/imaginary part addressing. For vector inputs, data operands can be permuted based on permutation indexes stored in Vector permutation register (VPR). Both the MMR and VPR can be pre-loaded during resource configurations or dynamically updated with values taken from the vector data bus. In addition to the permutation function, vector operands can be broadcast both horizontally and vertically to the SIMD core to support parallel computing, e.g., broadcasting $\underline{q}_i$ in 'Algorithm 1 line 10' to all processing lanes to compute multiple $\underline{r}_{i,k}$ in parallel.

The post-processing cell works with level-1, level-2, and accumulated (e-path) results from the SIMD core. It consists of two function units. With employed barrel shifters, the first unit is mainly used to dynamically adjust data precision of core-processing outputs. This is a useful function in vector processing especially for iterative and cumulative operations. Additionally, e-path output of each processing lane can be accumulated individually, which is required in supporting over-dimensioned vector operations (e.g., (6.4) for $N_{SW} > N$), where a folding technique performs data accumulations on partial data outputs. The second function unit provides capability of permuting vector outputs in ascending, descending, or user-defined order. For example, this feature can be used to perform sorting operations in MMSE-SQRD (Sect. 6.2.2).

6.4.3 Vector Data Memory Tile

Besides vector enhancements and multi-stage computation, the efficiency of the vector processor is contingent on memory access with regard to accessing bandwidth and flexibility. By inspection of algorithms discussed in Sect. 6.2, it is required that the SIMD core has access to multiple matrices and/or vectors in each operation, so as to avoid poor resource utilization and low throughput. As an example, efficient mapping of (Algorithm 1 line 3) requires two $N \times N$ matrix inputs, equivalent to having a $2 \times (4 \times 4) \times (16 + 16) = 1024$ bits/cycle memory bandwidth for a 16-bit 4×4 MIMO system. In addition to the bandwidth requirement, various forms of data accesses are needed, such as row- and column-wise addressing in matrix transposition. Moreover, to exploit additional DLP from independent data streams, accesses of vectors in different matrices are required by the multi-subcarrier processing. To meet these requirements, a hybrid memory organization and a flexible matrix access mechanism are adopted in the vector data memory tile (Tile-1).

Hybrid Memory Organization

To suffice the high memory accessing bandwidth, Tile-1 consists of vector and matrix access partitions, allowing simultaneous access of both vectors and matrices, see Fig. 6.25a. The basic element in both partitions is a dual-port memory cell, which provides a vector-level data storage and allows simultaneous read and write operations to ease memory access and improve processing throughput at the price of a larger memory footprint. In addition, the matrix partition provides direct matrix data access, which is realized by concurrently accessing a group of memory cells using only one set of address control. This arrangement is referred to as a memory page, shown in Fig. 6.25a. The vector accessing wordlength and the number of cells in a memory page are designed to match the processing capacity of the SIMD core in Tile-0, i.e., N scalar elements and N memory cells, respectively. On the other hand, the number of memory cells and pages are application dependent and should be optimized with respect to the bandwidth requirement and hardware cost. To ensure a sufficient memory storage required for the MIMO signal processing, Tile-1 in this study is deployed with two memory cells, for buffering data and storing R.MMSE-SW coefficients $\mathcal{W}$ in (6.4), and five pages, for storing $\hat{\underline{H}}$ (Fig. 6.26a) and $\underline{R}$. Details of these memory usages are further discussed in Sect. 6.5.

   Memory operations and accessing modes of each cell and page are managed by a local controller with configurations stored in a descriptor (DSC) table, see Fig. 6.25b. To communicate with other tiles, memory accesses are multiplexed using a crossbar network and interfaced through IO ports. For the array shown in Fig. 6.15b, Tile-1 contains four IO ports, allowing simultaneous access of two $N \times 1$ vectors and two $N \times N$ matrices for providing accesses to both Tile-0 and Tile-2. Referring to the aforementioned example, this corresponds to a memory bandwidth of 1280 bits/cycle.

Flexible Matrix Data Access

The presented multi-page memory arrangement and the crossbar network allow for the flexible data access required by the multi-subcarrier processing. For instance, by

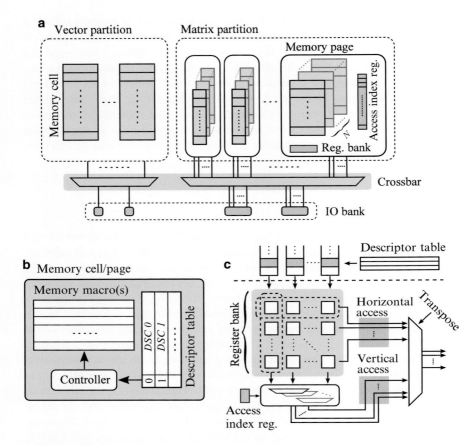

Fig. 6.25 Block diagram of the vector data memory tile (Tile-1), (**a**) a hybrid memory organization, (**b**) operation and accessing control, (**c**) data loading path of a memory page, supporting matrix access indexing and transposition

storing matrices of successive subcarriers in different memory pages, multiple data sets can be concurrently accessed and multiplexed based on arrangement indexes specified in memory configurations.

To better explain the necessity of the flexible memory access in supporting multi-subcarrier processing, the following shows a case study of various memory access patterns required for computing MMSE-SQRD (Algorithm 1). Figure 6.26a shows a memory layout of $\hat{\boldsymbol{H}}$ storage, where $\hat{\boldsymbol{H}}$ at even- and odd-indexed subcarriers (labeled as [0], [1], etc.) are stored in different memory pages. Because of the $2N \times N$ dimension, every $\hat{\boldsymbol{H}}$ is stored column-wise in two memory pages. In Fig. 6.26a, "$[0]c1a$" denotes column 1 of the upper half matrix ($N \times N$) of $\hat{\boldsymbol{H}}$ at subcarrier 0, and "$[0]c1b$" represents the lower half. Figure 6.26b shows a timing diagram of multi-subcarrier processing in Tile-0 for the computation of the first iteration

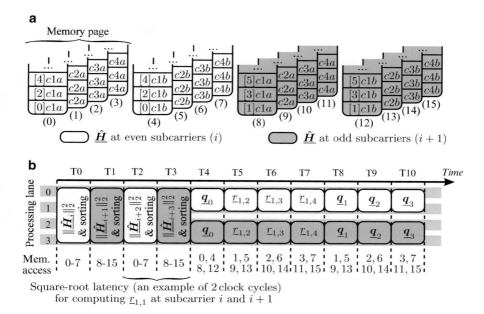

Fig. 6.26 Example of some memory access patterns required for computing MGS-based MMSE-SQRD in a 4 × 4 MIMO system. (**a**) Odd and even indexed $\hat{\boldsymbol{H}}$ are stored separately in different memory pages. (**b**) Timing diagram of the multi-subcarrier processing in the SIMD core with $N = 4$ processing lanes

of MMSE-SQRD. Together with the operations performed in each time interval, required memory access patterns are listed. For example, 0–7 indicates concurrent access of memory cells with index from (0) to (7). During the time interval T0–T3, multiple matrix accesses are needed for computing the squared Euclidean norm of $\hat{\boldsymbol{H}}$. Multiple vector readings from different memory pages are required during T4–T10. Supported by the flexible memory access schemes, multiple subcarriers can be processed in parallel to efficiently utilize processing gaps caused by data-dependent operations and computation latency. Without this support, processing lanes during the time intervals of shaded computations in Fig. 6.26b would be idle. Therefore, flexible memory access schemes are important for achieving high processing efficiency.

To further improve matrix access flexibility, a data arrangement circuit, illustrated in Fig. 6.25c, is implemented in each memory page. Specifically, data loaded from each memory page are buffered in a local register bank and are capable of being rearranged vector-wise in a vertical direction, based on an access index associated with each matrix storage. Benefiting from this setup, vector readouts from a matrix can be accessed freely in any order without physically exchanging data. This is useful, for example, in supporting sorted matrix accesses in MMSE-SQRD (Algorithm 1 line 5). The vector access indexes are stored in special registers that are transparent to the users and are configurable during every matrix data transfer. In addition to these index manipulations, the presented architecture is capable of outputting matrices in a transposed form (used, for example, in (6.12)) by selecting either the row or column output. As a result, processing cells are relieved from such data arrangement operations, which would otherwise result in enormous underused processing power. Moreover, physical data exchange and redundant memory accesses (due to read and write of the same data contents) are completely eliminated.

6.4.4   Scalar Resource Cells and Accelerators

In the scalar domain, Tile-2 and 3 perform scalar and conditional operations as well as dynamic configurations of other tiles in the array. Among them, Tile-2 consists of two scalar memories for storing data and configurations respectively. Tile-3 contains one memory for data buffering and three processing cells for computations. Figure 6.27 shows the three scalar processing cells in Tile-3, a generic signal processor and two acceleration units. The generic processor is a customized RISC with optimized conditional instructions and specialized functionality for dynamic RC configurations, similar to the one presented in Chap. 4. The two accelerators behave like co-processors of the generic processing cell for performing irregular operations, i.e., the inverse square root in MMSE-SQRD and the node perturbation in MMSE-NP, respectively.

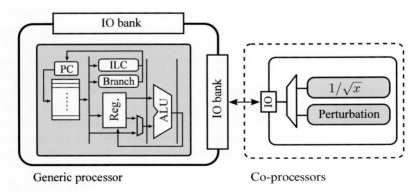

Fig. 6.27 Block diagram of the scalar processing cells in Tile-3, containing a generic RISC-structured processor and two accelerators

Inverse Square-Root Unit

To compute the inverse square root of x, where $x \in \mathbb{R}$ and $x > 0$, Newton's method is adopted in this study. Newton's method iteratively computes approximations to the root of a real-valued function $f(y)$ [16]. In the case of inverse square root, $f(y)$ is defined as,

$$f(y) = \frac{1}{y^2} - x, \qquad (6.22)$$

where $y = 1/\sqrt{x}$. A general expression of Newton's method for iteration i is written as

$$y_{i+1} = y_i - \frac{f(y_i)}{f'(y_i)}, \qquad (6.23)$$

where the whole process starts off with some arbitrary initial value y_0. Substituting (6.22) into (6.23), the output of each iteration can be expressed as

$$y_{i+1} = 2^{-1} y_i \left(3 - x y_i^2\right). \qquad (6.24)$$

After K iterations, the value of y_{i+1} converges to $1/\sqrt{x}$. The number of iterations required depends on the accuracy requirement of the application and how close the initial value y_0 is to $1/\sqrt{x}$. Fixed-point simulations show that $K = 2$ is sufficient in this study to obtain a near floating-point performance in terms of FER of the presented system setup (Sect. 6.3.1). It is worth mentioning that calculation of the square root can be obtained by multiplying the final result y_K by the input x.

Figure 6.28 shows the block diagram of the inverse square root unit, which is capable of computing both $1/\sqrt{x}$ and $\sqrt{x}$. It consists of three main building blocks,

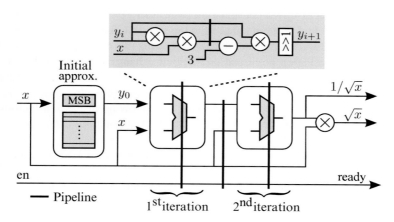

Fig. 6.28 Block diagram of the inverse square root unit using Newton' method with 2 iterations

an initial value approximation block and two function units. Given an input data x, the first block generates an initial value y_0 by looking up in a coefficient table. To reduce the table size while providing a good initial value, only three Most Significant Bits (MSBs) of x are used as inputs. The position of the MSB is dynamically detected for each input x. The basic principle of the adopted method is to share the same coefficients stored in the table for different initial value approximations. For example, $y_0 = 1/4$ for $x = 16$, where the coefficient $1/4$ ("01000000" in a unsigned radix-2 '0.8' format) can be used to generate the initial value for $x = 256$ by shifting it 2 bits to the right, that is, $y_0 = 1/16$ with the radix-2 representation of "00010000". Benefiting from the coefficient sharing, only 8 entries are required in the look-up table.

The two function units in Fig. 6.28 are used to compute (6.24), one per iteration. Each unit contains three real-valued multipliers, a subtracter, and a one-bit logical left shifter for realizing the 'divide by 2' operation. To increase the processing throughput, both units and their connections are pipelined, shown in the bold vertical lines in Fig. 6.28. Along with the square root and its reciprocal, a flow control signal 'ready' is provided on the output for indicating the computation status.

Node Perturbation Unit

In the adopted MIMO symbol detector MMSE-NP, one of the important steps is to find the nearest sibling symbols to the initial MMSE result x_p^{MMSE} (6.12) based on the criterion of (6.14). The entire perturbation process involves enormous fine-grained data manipulations, since x_p consists of normalized constellation points that are drawn from a finite set of integers, e.g., $\sqrt{42}x_{p(i)} \in [\pm 1, \pm 3, \pm 5, \pm 7]$ for 64-QAM. Therefore, the node perturbation process is implemented as an accelerator for attaining high implementation efficiency.

One way to find Ω_i closest symbols for each spatial layer is to compute distance between all possible M-QAM constellation points to $x_{p(i)}^{\mathrm{MMSE}}$, sort them in ascending order, and pick the first Ω_i points that have the smallest distance values. However, this brute-force method has high complexity, requiring two multiplications and three additions per constellation point and an M-point exhaustive search at the end. In contrast, this study adopts a Fast node enumeration (FNE) scheme [21], aiming to reduce the computational complexity by exploiting the geometric and symmetric properties of M-QAM. To better explain this, the following discussion focuses on 64-QAM and assumes $\max\{\Omega_i\} = 5$. Other system configurations can be processed by using the same concept.

Figure 6.29 illustrates the basic principle of the presented FNE scheme. The horizontal and vertical axis represent distance (δ) between the constellation points and the initial hard-output MMSE result $\hat{x}_{p(i)}^{\mathrm{MMSE}}$, for real and imaginary part, respectively. In this example, N_e is the closest symbol to $x_{p(i)}^{\mathrm{MMSE}}$, assuming $x_{p(i)}^{\mathrm{MMSE}}$ lies within the dashed box in Fig. 6.29. The distance between $x_{p(i)}^{\mathrm{MMSE}}$ and N_e is expressed as

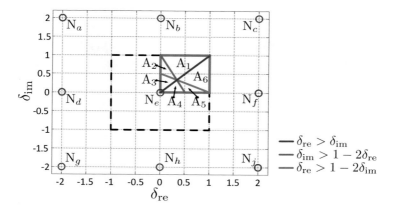

Fig. 6.29 Illustration of the fast node enumeration scheme

Table 6.7 Symbol sequences with respect to the position of $x_{p(i)}^{\text{MMSE}}$

Position of $x_{p(i)}^{\text{MMSE}}$	A_1	A_2	A_3	A_4	A_5	A_6
Symbol sequence	N_e	N_e	N_e	N_e	N_e	N_e
	N_b	N_b	N_b	N_f	N_f	N_f
	N_f	N_f	N_f	N_b	N_b	N_b
	N_c	N_d	N_d	N_h	N_h	N_c
	N_d	N_c	N_h	N_d	N_c	N_h

$$\delta = x_{p(i)}^{\text{MMSE}} - N_e = a + jb. \tag{6.25}$$

Utilizing the symmetric and equidistant property of M-QAM, a and b in (6.25) can be shifted around N_e to reduce the search space to the first quadrant. As a result, $\{a, b\} \in [0, 1]$. To find the remaining neighboring symbols, distances between $x_{p(i)}^{\text{MMSE}}$ and other constellation points in Fig. 6.29 are computed and sorted in ascending order. By analyzing the resulting order of those symbols with respect to the position of $x_{p(i)}^{\text{MMSE}}$, the search space can be divided into six unique zones that cover all possible symbol sequences. These zones are labeled with A_1 to A_6, see Fig. 6.29. The corresponding symbol sequences are listed in Table 6.7.

To determine the zone in which $x_{p(i)}^{\text{MMSE}}$ resides, the real and imaginary value of δ in (6.25) are compared using the following criteria:

1. $\delta_{\text{re}} > \delta_{\text{im}}$,
2. $\delta_{\text{im}} > 1 - 2\delta_{\text{re}}$,
3. $\delta_{\text{re}} > 1 - 2\delta_{\text{im}}$.

These comparisons correspond to three boundary lines inside the dashed box in Fig. 6.29. Once $x_{p(i)}^{\text{MMSE}}$ is positioned, all the required nearest sibling symbols are obtained. This is realized with the help of a look-up table, which stores all symbol sequences listed in Table 6.7 with different boundary check. Worth mentioning is

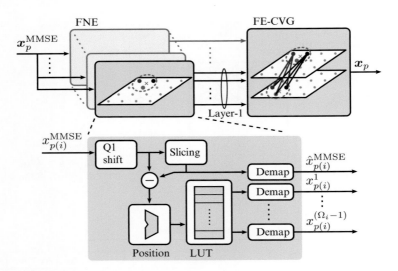

Fig. 6.30 Block diagram of the node perturbation unit

that the adopted FNE scheme can be applied to other cases, where the search space is not at the center of the constellation map, for example, at corners or borders [21].

Figure 6.30 shows the block diagram of the node perturbation unit. It consists of N FNE units, one for each spatial layer, and a candidate vector generation unit for constructing candidate vectors by using the FE-CVG method (6.2.3). The FNE process starts by shifting the input $x_{p(i)}^{\mathrm{MMSE}}$ into the first quadrant. The initial hard-output symbol $\hat{x}_{p(i)}^{\mathrm{MMSE}}$, i.e., the closest symbol, is found by slicing $x_{p(i)}^{\mathrm{MMSE}}$ to the nearest constellation point. After calculating the distance (δ) between $\hat{x}_{p(i)}^{\mathrm{MMSE}}$ and $x_{p(i)}^{\mathrm{MMSE}}$, the position block carries out all comparisons of δ in parallel to determine the position of $x_{p(i)}^{\mathrm{MMSE}}$. The resulting zone number, ranging from A_1 to A_6, is used as an input to the following Look-up table (LUT) to obtain the remaining symbols. The FNE process is completed by shifting the expanded symbols back to the original quadrant, performed by de-mapping units.

6.4.5 Concurrent Candidate Evaluation

In this section, a technique to further improve the implementation efficiency is presented. Among the MIMO signal processing, the ED calculation ($\|\cdot\|$) in (6.16) is the most compute-intensive operation, which needs to be performed at every data-carrying subcarrier and for each of the L candidate vectors, e.g., $L = 24$ for $\Omega = [\mathrm{F}, 4, 3, 2]$. A straightforward mapping of this on the SIMD core tends to incur low hardware utilization, at most 50 % when computing $\underline{R}x_p$, since $\underline{R}$ is an upper triangular matrix with real-valued diagonal elements. This is impermissible from the hardware efficiency point of view. To tackle this problem, the property of $\underline{R}$ is utilized in such a way that two candidate vectors are concurrently evaluated, with the second $\underline{R}x_p$ operation mapped to the lower triangular part of the SIMD core. In the following, computation of $\underline{R}x_p$ in a 2×2 MIMO system using a 2×2 SIMD core is given as an example to better illustrate the concurrent candidate evaluation.

To fully utilize the $N \times N$ CMAC units in the SIMD core, $\underline{R}$ is duplicated to process two different x_p vectors at the same time. This matrix duplication is achieved by mirroring the $\underline{R}$ matrix in both vertical and horizontal directions, such that the two matrices ($\underline{R}$ and its counterpart $\underline{R}'$) together compose a full square matrix, as illustrated at the bottom of Fig. 6.31. The required input vectors x_{p_0} and x_{p_1} are fed to the SIMD core via the matrix path MA and the vector path VC,

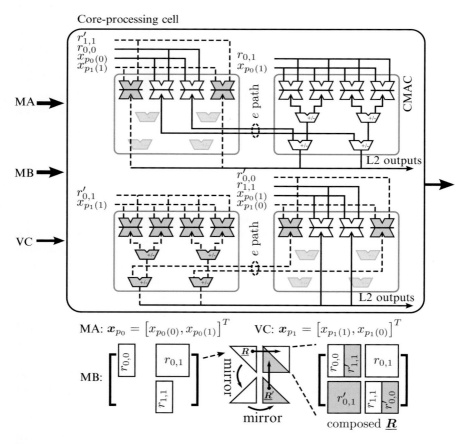

Fig. 6.31 Concurrent candidate evaluation using the SIMD-based core-processing cell, an example of a 2×2 MIMO system. Internal data multiplexers are omitted for simplicity. Shaded blocks and dashed lines illustrate the computation of the second $\underline{R}x_p$ operation, denoted as $\underline{R}'x_{p_1}$

respectively. Note that reverse-order permutation is required for the vector input x_{p_1} to match with the matrix orientation of $\underline{R}'$ (flipped upside down). Figure 6.31 illustrates input data arrangement and internal data processing flow of the SIMD core. The ones associated with $\underline{R}'x_{p_1}$ computation are depicted in shaded blocks and dashed lines. In the diagonal CMAC units, processing of both x_{p_0} and x_{p_1} co-exist because of the real-valued diagonal elements in $\underline{R}$. Additionally, both level-1 and level-2 adders are bypassed in these CMACs, shaded in light grey in Fig. 6.31, since only real-valued multiplications are performed. Final vector outputs of both computations are conveyed to the following processing cell via level-2 outputs. As a result of this concurrent candidate evaluation, both hardware utilization and processing throughput are doubled for $\underline{R}x_p$ computations.

It is worth mentioning that various techniques have been presented in literature for improving the hardware efficiency of ED computations. For example, utilizing the property of the candidate vectors x_p (i.e., constellation points), [34, 52] simplify the computation of $\underline{R}x_p$ by performing finite alphabet multiplications. However, applying those accelerator-based design techniques on vector processors may be infeasible or cost ineffective, as they require either fine-grained data manipulations (e.g., to realize finite alphabet multiplications) or increased data path width (e.g., to hold multiple $\underline{R}x_p$ outputs). In contrast, the adopted scheme utilizes the existing structure of the SIMD core and only requires a few specialized signals for controlling diagonal CMAC units. Thus, it provides a balance between hardware efficiency and complexity.

6.5   Implementation Results and Comparison

To cope with different system configurations and design constraints on, for example, antenna size and processing throughput, the heterogeneous cell array is fully parameterizable at system design-time. Figure 6.32 shows the detailed architecture of the cell array configured for the target 20 MHz 4×4 MIMO LTE-A downlink. Processing and memory cells in the vector domain are labeled with 'VPC-$\mathcal{X}$' and 'VMC-$\mathcal{X}$' respectively, while those in the scalar domain are denoted as 'SPC-$\mathcal{X}$' and 'SMC-$\mathcal{X}$'. All data computations are performed in 16 bits fixed-point arithmetic with 8 guard bits for accumulations. In Tile-0, the core-processing cell is configured to have 4×4 CMAC units. The post-processing cell contains two 3-bit barrel shifters deployed for handling data from matrix and vector bus respectively. The register bank consists of 16 general-purpose vector registers, 16 VPRs, and 16 MMRs. Each distributed configuration memory can store up to 16 hardware configurations, and the program memory is capable of storing 256 micro-code sequences.

The generic processor in Tile-3 is configured to have a 3-bit barrel shifter, 11-bit one-level inner loop controller, 16 general-purpose scalar registers, and a 18 Kb program memory. The node perturbation unit in SPC-1 is able to extend each symbol with up to five nearest neighbors and generate one candidate vector in every clock cycle.

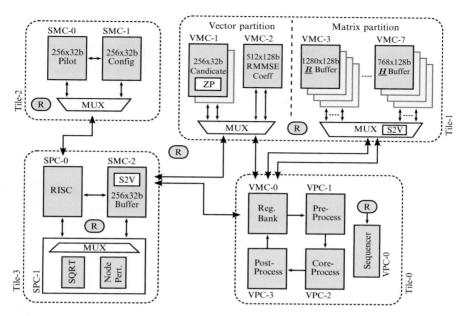

Fig. 6.32 Architecture diagram of the heterogeneous cell array configured for baseband processing in a 20 MHz 4×4 MIMO LTE-A downlink

As for memory cells, each is configured to have 4 DSCs. In VMC-2, SMC-0, and SMC-2, the micro-block function (Sect. 5.3.2) is enabled. Data memories are deployed mainly to suffice the storage requirement of the target LTE-A setup. In the current design, the array contains 2.34 Mb of memory, in which 88 % are data buffers for keeping data required in one LTE-A time slot (e.g., channel and decomposed matrices), 2 % are control memories for storing instructions and resource configurations, and 10 % are reserved space for facilitating flexible algorithm mappings and future system updates. Detailed memory configurations of all RCs are summarized in Table 6.8. Data transfers from vector to scalar RCs are bridged by memory cells using the micro-block function, whereas the reverse paths are handled by dedicated scalar-to-vector adapters, shown as 'S2V' in Fig. 6.32. Each adapter contains a vector register and a FSM, capable of transmitting N scalar data packets in a vector form to the receiving end. Moreover, it should be mentioned that VMC-1 in Tile-1 is dedicated to storing candidate vectors x_p in symbol detection. Therefore, the wordlength of the memory is substantially reduced by only storing M-QAM values. For 64-QAM modulation, each symbol in x_p requires 2×4 bits instead of 2×16, reducing the memory requirements by 4 times. During memory reading, the M-QAM values are extended to the vector format by padding zeros. This QAM-to-vector converter is denoted as 'ZP' in Fig. 6.32.

Table 6.8 Memory configurations and usages

Resource cell		Memory		Reserved	Usage
Tile-0	VPC-0	256 × 32b	8 Kb	N/A[a]	Program memory
	VPC-1–3 VMC-0	(16 × 32b) × 7	3.5 Kb	N/A[a]	Configuration memory
Tile-1	VMC-1	(256 × 32b) × 2	16 Kb	43.75 %	Candidate vector buffer
	VMC-2	512 × 128b	64 Kb	43.75 %	R.MMSE-SW $\mathcal{W}$ ROM
	VMC-3	(1280 × 128b) × 4	640 Kb	6.25 %	$\boldsymbol{R}$ buffer
		1280 × 8b	10 Kb	6.25 %	Access index
	VMC-4–7	(768 × 128b) × 16	1536 Kb	8.85 %	$\boldsymbol{H}$ buffer
		(768 × 8b) × 4	24 Kb	8.85 %	Access index
Tile-2	SMC-0	256 × 32b	8 Kb	42.38 %	Pilot ROM & data buffer
	SMC-1	256 × 32b	8 Kb	N/A[a]	Configuration memory
Tile-3	SPC-0	384 × 48b	18 Kb	N/A[a]	Program memory
	SPC-1	N/A	N/A	N/A	N/A
	SMC-2	256 × 32b	8 Kb	N/A[a]	Data buffer
Network		N/A	N/A	N/A	N/A

[a] Reserved space is not assessed for control and configuration memories

6.5.1   Implementation Results

The cell array is modeled in VHDL, synthesized using Synopsys Design Compiler with a 65 nm CMOS standard digital cell library, and routed using Cadence SoC Encounter. Counting a two-input NAND gate as one equivalent gate, the whole array contains 2.76 M gates and has a core area of 8.88 mm^2 at 74 % cell density in chip layout. Data buffers (Table 6.8) occupy more than 60 % of the area, while the logic blocks, including control memories and the hierarchical network, share the rest. Excluding those data buffers, it shows in Table 6.9 that most of the logic gates are devoted to the vector processing domain (i.e., Tile-0 and 1) and the on-chip network takes less than 5 %. At 1.2 V nominal core voltage supply, a maximum clock frequency of 500 MHz is obtained from post-layout simulations with back annotated timing information. At this frequency, the array is capable of performing 8.5 G CMACs per second, considering the 4 × 4 CMAC bank in Tile-0 and the CMAC unit in the generic processor in Tile-3.

Vector Dataflow Processor

Figure 6.33a shows the area breakdown of Tile-0, the vector dataflow processor. The control logic of the processor, including the sequencer and the program memory, occupy 22 % of the total area, while the processing cells take 51 % and the register bank consumes 27 %. Note that the distributed configuration memories are counted as part of the sequencing control in Fig. 6.33a. The relatively low area consumption

Table 6.9 Area and power breakdown of the cell array without data buffers

Resource cell		Gate count [KG]		Power [mW]	
Tile-0		367	34.77 %	164.93	53.75 %
Tile-1	Vector partition	96	9.12 %	5.99	1.95 %
	Matrix partition	365	34.60 %	68.06	22.18 %
Tile-2		47	4.44 %	3.20	1.04 %
Tile-3	Generic processor	70	6.60 %	44.10	14.37 %
	Others	61	5.83 %	16.98	5.53 %
Network		49	4.65 %	3.56	1.16 %
Total		1055	100.00 %	306.84	100.00 %

of the control logic reveals the low control overhead of the processor, thanks to the adopted distributed micro-code execution scheme. Among the multi-stage computation path, the core-processing cell consumes most of the area due to the deployed homogeneous CMAC bank.

Vector Data Memory Tile

Tile-1 consists of two memory cells, VMC-1 to 2, and five memory pages, VMC-3 to 7. Because of the large storage requirements of the application, e.g., to store channel matrices $\underline{H}$ and $\underline{R}$ for all 1200 subcarriers and coefficients $\mathcal{W}$ for the R.MMSE-SW estimator, most of the area in Tile-1 is consumed by memory macros, see Fig. 6.33b. Among the control logic, memory pages consume 80 % of the area, due to the employed flexible access schemes such as matrix data transposition and access indexing. It is worth mentioning that VMC-2 is configured to have the micro-block function used to interface with RCs in the scalar processing domain. Therefore, it can be seen from Fig. 6.33b that the control logic of VMC-2 consumes slightly more area than that of the VMC-1.

Resource Cells in Scalar Processing Domain

In the scalar processing domain, the generic processor and accelerators occupy around 45 % of the area, see Fig. 6.33c. Compared to the vector dataflow processor in Tile-0, the generic processor is equipped with a larger program memory and a simpler data path, since it is designed to mainly perform control related operations, such as conditional instruction execution and configuration of other RCs. This can be seen from the area partition in Fig. 6.33c, where the control logic of the processor takes almost the same area as its logic part, i.e., 11.83 % versus 14.27 %.

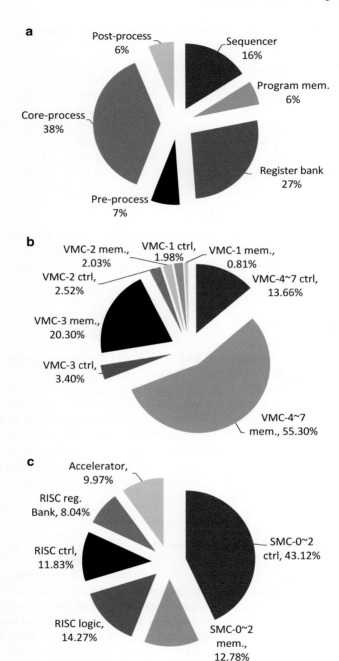

Fig. 6.33 Area breakdown of RCs in the reconfigurable cell array. (**a**) Vector dataflow processor (Tile-0). (**b**) Vector data memory tile (Tile-1). (**c**) Tile-2 and 3 in the scalar processing domain

6.5.2 Task Mapping and Timing Analysis

The MIMO processing tasks, i.e., channel estimation, pre-processing, and symbol detection, are manually mapped onto the cell array with a primary focus on sufficing the stringent timing constraint and achieving high processing throughput. To this end, multi-subcarrier processing is adopted in all tasks and is scheduled based on the LTE-A resource block, i.e., 12 consecutive subcarriers. The number of blocks to process in each computation step is determined manually based on the computation and communication latency and available hardware resources. For example, MMSE-SQRD is programmed to operate on 2 LTE-A resource blocks in each step due to its high data dependency and the long latency involved in obtaining results from the inverse square root unit. In contrast, the other two tasks work with 1 resource block at a time. In addition to the multi-subcarrier processing, most of the data transfers are scheduled to utilize the low-latency high-bandwidth local interconnects, while the hierarchical network is mainly used for resource configurations and the streaming of external data such as receiving vector y and decoded $\hat{x}$. In the following, detailed mapping of the MIMO processing tasks is described.

Channel Estimation

Recall that channel estimation contains two computation steps, LS estimation at pilot tones and H interpolation for data-carrying subcarriers. In this study, the LS computation is performed by the generic processor in Tile-3 and results are stored in an LS buffer (VMC4–7) in Tile-1. The computation starts immediately data at pilot positions have been received. Data transfers between SPC-0 and memory pages in Tile-1 are carried out through the hierarchical network. In Tile-1, the received scalar data packets are converted to the vector format before writing to the memory. This is accomplished by using the S2V unit deployed in the matrix partition of Tile-1. Steps ① to ⑦ in Fig. 6.34 illustrate the processing flow of the LS computation. Using LS estimated channel coefficients, the data-tone H interpolation is performed by the vector dataflow processor. To fully utilize the 4×4 CMAC bank, four Rx spatial layers (i.e., rows of H) are computed in parallel, one per processing lane, in each clock cycle. For $N_{SW} = 24$, each interpolation process requires $24/4 = 6$ iterations to complete. The intermediate results are accumulated in the post-processing cell in Tile-0. Since the pilot tones reside in every third subcarrier in an OFDM symbol (Fig. 6.2), processing of each LTE-A resource block is divided into four groups, each containing one pilot and two data tones. Table 6.10 shows the pseudo-code of the data-tone H interpolation. Taking advantage of the adopted multi-level zero-delay inner loop controller, loop operations are used whenever possible, aiming to attain a modular program structure and to ease parameter updates.

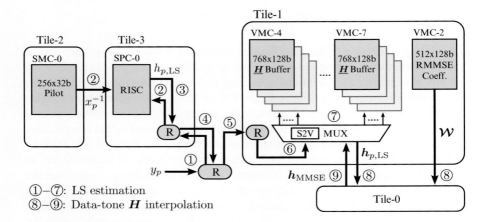

Fig. 6.34 Processing flow of the channel estimation, performed by the generic processor in Tile-3 and the vector dataflow processor (Tile-0)

Table 6.10 Pseudo-code of the data-tone H interpolation performed in Tile-0

for $i = 1 : 100$ **do**	% Loop for 100 LTE-A resource blocks
for $j = 1 : 4$ **do**	% 4 groups per LTE-A resource blocks
Copy H @ pilot position from "LS buffer" to "$\underline{H}$ buffer"	
for $k = 1 : 2$ **do**	% 2 data tones per group per resource block
for $l = 1 : 6$ **do**	% 6 iterations per data tone for $N_{SW} = 24$
for $n = 1 : 4$ **do**	% 4 Tx spatial layers, i.e., 4 columns of H
'VDP$(ab + e)$' to compute $h_{\text{MMSE}} = \mathcal{W}h_{p,\text{LS}}$	
end for	
end for	
end for	
end for	
end for	

Channel Matrix Pre-processing

The MMSE-SQRD based channel matrix pre-processing is mainly performed by the vector dataflow processor, except that $1/\sqrt{x}$ and $\sqrt{x}$ operations are outsourced to the inverse square root unit in Tile-3. Taking the augmented channel matrix $\underline{H}$ as an input, the MGS-based MMSE-SQRD performs matrix orthogonalization iteratively based on Algorithm 1. The corresponding task mapping on the vector dataflow processor and the layout of data storage in Tile-1 are illustrated in Fig. 6.26. It should be pointed out that, among numerous task mapping schemes, the adopted approach focuses on the utilization of hardware resources in Tile-0 and the modularity of the program structure. Briefly, the processor works with one subcarrier in each clock cycle during the norm computation of $\underline{Q}$, whereas

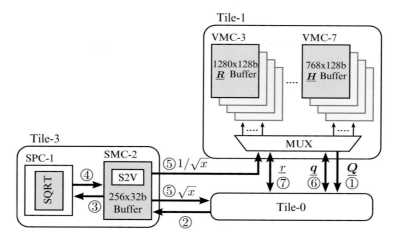

Fig. 6.35 Processing flow of the MMSE-SQRD based channel matrix pre-processing performed by the vector dataflow processor (Tile-0)

two subcarriers are processed in parallel during other operations, i.e., column vector updates in $\underline{Q}$ and the computation of $\underline{R}$. Upon the completion of the matrix orthogonalization process, $\underline{H}^{-1} = 1/\sigma_n \underline{Q}_b \underline{Q}_a^H$ is computed as a post-processing, required in the following symbol detection (6.12). Considering the latency of the inverse square root operation, i.e., 3 clock cycles computation latency (Sect. 6.4.4) plus 4 clock cycles communication latency illustrated in Fig. 6.35, two resource blocks are processed in each computation step of MMSE-SQRD. Table 6.11 shows the pseudo-code of the task mapping. Note that column permutations of matrices $\underline{Q}$ and $\underline{R}$ are realized by manipulating the accessing indexes (Sect. 6.4.3) of matrix pages in Tile-1, thus requiring no physical data exchanging.

Symbol Detection

The MMSE-NP based symbol detection contains three main computation steps: initial MMSE detection, symbol expansion and candidate vector generation, and candidate evaluation. Among these, the second operation is performed by the node perturbation unit in Tile-3, while the other two are handled by the vector dataflow processor. Table 6.12 shows the pseudo-code of the computations performed in the vector processor. In view of the latency of communication between the vector processor and the node perturbation unit in Tile-3, see Fig. 6.36, 12 adjacent data-carrying subcarriers are processed in each computation step of MMSE-NP. For the target 20 MHz LTE-A, each time slot has 7200 data-carrying subcarriers (Table 6.4), resulting in $7200/12 = 450$ blocks to be processed. To achieve high utilization of the CMAC bank in Tile-0, each ED computation (6.16) during the evaluation of

Table 6.11 Pseudo-code of MMSE-SQRD, computations performed in Tile-0

---

for $i = 1 : 50$ **do**   % *Loop for 50 pairs of LTE-A resource blocks*
  for $j = 1 : 4$ **do**   % *4 columns of $\underline{H}$*
    for $k = 1 : 2 \times 12$ **do**   % *Processing one subcarrier per clock cycle*
      'VDP$(ab + e)$' to compute $\boldsymbol{\xi} = \left[\|\underline{\boldsymbol{q}}_{j+1}\|_2^2, \|\underline{\boldsymbol{q}}_{j+2}\|_2^2, \ldots, \|\underline{\boldsymbol{q}}_N\|_2^2 \right]^T$
      Sorting to obtain $\boldsymbol{\xi}_{min}$
    end for
    for $k = 1 : 12$ **do**   % *Processing two subcarriers per clock cycle*
      'bc' to compute $\underline{\boldsymbol{q}}_j = \underline{\boldsymbol{q}}_j / r_{j,j}$
    end for
    for $k = 1 : 12$ **do**   % *Processing two subcarriers per clock cycle*
      for $l = j + 1 : 4$ **do**
        'VDP$(ab + e)$' to compute $r_{j,l} = \underline{\boldsymbol{q}}_j^H \underline{\boldsymbol{q}}_l$
      end for
    end for
    for $k = 1 : 12$ **do**   % *Processing two subcarriers per clock cycle*
      for $l = j + 1 : 4$ **do**
        '$a - bc$' to compute $\underline{\boldsymbol{q}}_l = \underline{\boldsymbol{q}}_l - r_{j,l} \underline{\boldsymbol{q}}_j$
      end for
    end for
  end for
  for $j = 1 : 2 \times 12$ **do**   % *Post-processing*
    'VDP$(ab + e)$' & 'bc' to compute $\underline{H}^{-1} = 1/\sigma_n \boldsymbol{Q}_b \boldsymbol{Q}_a^H$
  end for
end for

---

candidate vectors is performed in three sub-steps: $\boldsymbol{\alpha} = \underline{\boldsymbol{R}} \boldsymbol{x}_p$, $\boldsymbol{\beta} = \tilde{\boldsymbol{y}} - \boldsymbol{\alpha}$, and vector norm $\|\boldsymbol{\beta}\|_2^2$. In each clock cycle, the computation of $\underline{\boldsymbol{R}} \boldsymbol{x}_p$ operates on two candidate vectors by using the concurrent candidate evaluation scheme (Sect. 6.4.5), while the other two vector-based operations work with four candidates at a time. Upon the completion of each vector norm computation $\|\boldsymbol{\beta}\|_2^2$, the four candidate vectors under evaluation are sorted in the post-processing cell in Tile-0. The one with the smallest ED value is temporarily stored in the register bank for further comparisons with other candidates. For $\boldsymbol{\Omega} = [F, 4, 3, 2]$, there are in total 24 candidate vectors in each symbol detection, implying $24/4 = 6$ temporarily stored candidates at the end of the ED computation. The final detection output is obtained by comparing these 6 candidates, finding the one with the smallest value, and loading the corresponding candidate vector from VMC-1 together with the permutation matrix ($\boldsymbol{P}$ in (6.5)) from VMC4. Finally, the recovered transmitted vector $\hat{\boldsymbol{x}}$ with its original symbol sequence is sent out through the hierarchical network.

Table 6.12 Pseudo-code of symbol detection, computations performed in Tile-0

---

for $i = 1 : 7200/12$ **do**   % *Loop for 7200/12 blocks*
  for $j = 1 : 12$ **do**   % *Loop for 12 subcarriers*
    'VDP$(ab + e)$' to compute $x_p^{\text{MMSE}} = \underline{H}^{-1}y$
  end for
  for $j = 1 : 12$ **do**
    'VDP$(ab + e)$' to compute $\tilde{y} = Q_a^H y$
  end for
  for $j = 1 : 12$ **do**
    for $k = 1 : 24/2$ **do**   % *Processing two $\underline{R}x_p$ per clock cycle*
      'VDP$(ab + e)$' to compute $\underline{R}x_p$
    end for
  end for
  for $j = 1 : 12$ **do**
    for $k = 1 : 24/4$ **do**   % *Processing four $\tilde{y} - \underline{R}x_p$ per clock cycle*
      '$a - b$' to compute $\tilde{y} - \underline{R}x_p$
    end for
  end for
  for $j = 1 : 12$ **do**
    for $k = 1 : 24/4$ **do**   % *Processing four $\|\cdot\|_2^2$ per clock cycle*
      'VDP$(ab + e)$' to compute $\left\| \tilde{y} - \underline{R}x_p \right\|_2^2$ & sorting
    end for
  end for
  for $j = 1 : 12$ **do**
    for $k = 1 : \lceil 24/4/4 \rceil$ **do**   % *Post-processing*
      Sorting to find $\hat{x}_p = \arg \min_{x_p \in S} \left\| \tilde{y} - \underline{R}x_p \right\|_2^2$
    end for
    $\hat{x} = P\hat{x}_p$ % *Final detection output*
  end for
end for

Miscellaneous Operations

Besides the aforementioned MIMO processing tasks, various miscellaneous operations are required, e.g., memory initialization for the permutation matrix P (6.5) and the augmented channel matrix $\hat{\underline{H}}$ (6.6). These operations occupy only a fraction of the total processing time and are performed in the beginning of each processing iteration (i.e., time slot).

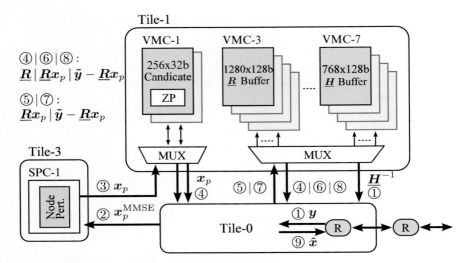

Fig. 6.36 Processing flow of the MMSE-NP based symbol detection performed by the vector dataflow processor (Tile-0)

Table 6.13 Overhead analysis for computing MIMO processing tasks

	Execution time [Clock cycle]	Control time [Clock cycle]	Control overhead [%]
Ch. estimation	20,801	1201	5.77
QRD	22,451	851	3.79
QRD[a]	15,101	701	4.64
Detection	190,201	3001	1.58
Total	232,203	5003	2.16

a Without post-processing

Results and Discussions

To assess the efficiency of the task mapping, control overhead is analyzed for program executions on the vector dataflow processor. Here the control is defined as non-computational operations, such as loop initialization and run-time program updates. Table 6.13 lists the total execution time and the number of control operations required for accomplishing three MIMO processing tasks. As can be seen, the total control overhead measured on the vector dataflow processor is only about 2 % of the total execution time, thanks to the algorithm–architecture co-design. On the algorithm side, benefiting from the adopted streaming-based processing flow in each individual task, branch operations are completely eliminated, see Tables 6.10, 6.11, and 6.12. On the architecture side, with the employed configurable ISA, the number of hardware configurations (micro-codes) and program updates is substantially reduced, since the data path of the processor can be dynamically configured to better

Table 6.14 Performance summary of the MIMO signal processing

	Clock Cycle/Op	Time [μs]	Throughput	Power[b] [mW]	Energy[b]
Ch. estimation	17.33	41.60	28.84 MEst/s	276.24	9.58 nJ/Est
QRD	18.71	44.90	26.72 MQRD/s	314.05	11.75 nJ/QRD
QRD[a]	12.63	30.30	39.60 MQRD/s	315.36	7.96 nJ/QRD
Detection	26.42	380.40	454.26 Mb/s	280.82	0.62 nJ/b
Miscellaneous	2.34	2.82	N/A	269.99	0.81 nJ/op
Total/average	32.62	469.72	367.88 Mb/s	306.84	0.83 nJ/b

[a] Without post-processing
[b] With data buffers excluded

suit target applications. Additionally, loop operations are assisted by the multi-level zero-delay inner loop controller, thus requiring no loop manipulations from the user.

Table 6.14 summaries achieved performance of the three task mappings. Operating at 500 MHz, the total processing time for one LTE-A time slot is 469.72 μs. This fulfills the real-time requirement of the target LTE-A setup, i.e., $t_{iter} \leq t_{slot} = 0.5$ mS (see Sect. 6.3.4), and results in about 6 % spare time that can be used to map more advanced algorithms or upgrade system parameters such as the Ω assignment in symbol detection. Based on the processing time and the number of tones/bits required to compute, Table 6.14 presents the corresponding throughput achieved in each task. On average, recovering one transmitted vector $\hat{x}$, with all three processing tasks involved, requires 32.62 clock cycles, which is equivalent to a throughput of 367.88 Mb/s.

6.5.3 Computation Efficiency

To evaluate the computation efficiency of the array, resource utilization of the SIMD core in Tile-0 is measured as a representative, since it contributes to more than 90 % of the total computation capacity. Thanks to the vector enhanced SIMD structure (Sect. 6.4.2) and the multi-stage computation chain (Sect. 6.4.2), an average utilization of 77 % is achieved during the whole MIMO signal processing. Figure 6.37 reports a detailed utilization graph during the computation of two LTE-A resource blocks. Among these tasks, the miscellaneous processing shows the lowest utilization value, as the vector processor during that time interval only performs simple operations, such as vector scaling and data masking used for initializing registers and memory cells.

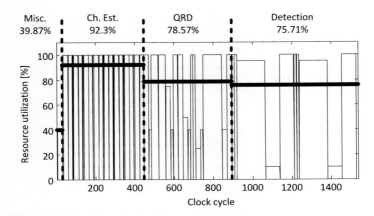

Fig. 6.37 Utilization of the SIMD core in Tile-0 during MIMO signal processing of two LTE-A resource blocks (24 subcarriers). *Horizontal lines* in the figure show the average utilization of the corresponding task

6.5.4 Power and Energy Consumption

Power consumption of the cell array is obtained from Synopsys PrimeTime using the post-layout design annotated with switching activities. At 500 MHz with 1.2 V supply voltage, the average power consumption for processing one data-carrying tone is 548.78 mW, including 306.84 mW from logic blocks and 241.94 mW from data buffers. The corresponding energy consumption for processing one information bit is 0.83 nJ/b and 1.49 nJ/b, without and with data buffers respectively. Table 6.14 summaries average power and energy consumption of different tasks with data buffers excluded. As can be seen, power consumption of different task mappings is quite balanced because of the high computation efficiency achieved by the algorithm–architecture co-design.

To acquire a more comprehensive understanding of the power distribution, a tile-level power breakdown of the array is listed in Table 6.9 and plotted in a pie-diagram in Fig. 6.38a. Among all, Tile-0 is the most power consuming block, because of the large area occupation and high resource utilization throughout the whole processing period. Further, the power distribution in Tile-0 is presented in Fig. 6.38b. The SIMD core accounts for 50 % of the power. The register bank consumes only 16 % thanks to that the developed multi-stage computation chain substantially reduces intermediate result buffering. Moreover, the hierarchical network of the array consumes only ~1% of the total power (Fig. 6.38a), due to the high allocation of local data transfers. Worth mentioning is that no special low power design techniques, such as clock and power gating and multiple power islands, are adopted in the current array. Therefore, further power savings are possible with a more advanced back-end design. Additionally, it should be pointed out that simulated power figures from the post-layout design may be different from chip measurement results.

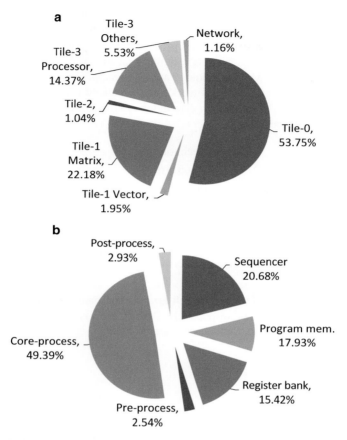

Fig. 6.38 Power breakdown of (**a**) the reconfigurable cell array and (**b**) the vector dataflow processor (Tile-0)

6.5.5   Comparison and Discussion

In this section, implementation results and various performance metrics of the cell array are summarized and compared with previously reported designs in open literature. In fact, a fair quantitative comparison with related work is not an easy task due to many different design factors, such as flexibility, algorithm selection, performance, and operating scenario. Which platform/technology to choose is highly dependent on design specifications such as system setup, area and power budget, Quality of service (QoS), and scalability requirements. Therefore, the following discussion only serves to give an overview of the design efficiency for related implementations and aims to position the cell array with respect to different performance measures. To ease the discussion, related hardware architectures are divided

into three broad categories: task specific accelerators (ASICs), programmable and
reconfigurable platforms (e.g., DSPs, FPGAs, and GPUs), and domain-specific
reconfigurable platforms (e.g., baseband processors).

To ensure a fair enough comparison, technology scaling is considered. Taking the
employed setup "65 nm CMOS technology and 1.2 V supply voltage" as a reference,
implementation results of all related work are normalized using a process scaling
factor s_{CMOS} [42]. The definition of s_{CMOS} and the normalization of frequency, area,
and power consumption are

$$s_{CMOS} = \frac{\text{Technology}}{65\,\text{nm}},$$

$$\text{Frequency}_{norm} \approx \text{Frequency} \times s_{CMOS},$$

$$\text{Area}_{norm} \approx \text{Area} \times \frac{1}{s_{CMOS}^2}, \tag{6.26}$$

$$\text{Power}_{norm} \approx \text{Power} \times \frac{1}{s_{CMOS}} \left(\frac{1.2\,\text{V}}{\text{Voltage}}\right)^2.$$

In Tables 6.15, 6.16, and 6.17, the cell array is compared with three aforementioned architecture categories, respectively. Performance is evaluated by assessing
area and power efficiency as well as hardware flexibility.

Area Efficiency

Area efficiency is calculated by normalizing the throughput of each processing task
to the corresponding hardware consumption. The presented solution accomplishes
three tasks within the tight timing constraint of the 20 MHz 4 × 4 MIMO 64-
QAM LTE-A downlink, thanks to the algorithm–architecture co-design, which
has more than 98 % of the total operations mapped onto the vector processor for
exploiting extensive DLP and attaining high resource sharing. Compared to other
implementations in Table 6.15, which adopt either lower-dimensions of MIMO
configurations or mapping of a single task, the cell array achieves the highest
throughput and shows superior area efficiency. Note that [40] and [32] in Table 6.15
are employed in single-antenna systems, Digital video broadcasting (DVB) and
Wideband code division multiple access (WCDMA), respectively. According to [5],
the performance required for WCDMA is less than 10 % of the one needed for
4×2 MIMO 3GPP LTE delivering 10 Mbps. Thereby, their results cannot be directly
compared with those of the baseband processing in MIMO systems. Here, they are
included for references only.

Compared to programmable and reconfigurable platforms in Table 6.16, the
processing throughput of the cell array is 2.8 and 45 times higher than that of
the FPGA and the DSP solution [26], respectively. Besides, its area efficiency
outperforms the GPU and CPU approaches by 3–5 orders of magnitude. It is

Table 6.15 Comparison of the cell array with reconfigurable baseband processors

		09 [40]	10 [32]	10 [39]	08 [28]	11 [29]	10 [11]	09 [12]	This work	
Antenna		–	–	4 × 4	2×2	4 × 4	4×2	2×2	4 × 4	
Modulation (QAM)		64	N/A	–	64	64	N/A	N/A	64	
Mapping (CE|QRD|DT)		√|–|√	√|–|√	–|√|–	–|–|√	–|–|√	√|√|√	√|√|√	√|√|√	
Technology [nm]		120	130	65	90	130	65	90	65	
Area [mm²]		11	11	N/A	N/A	N/A	16.06ᶜ	32	8.88	
Gate count [KG]		200[a]	N/A	824[a]	1200	71[a]	5969[c]	N/A	2760	1055[a]
Frequency [MHz]		70	300	234	600	277	400	400	500	
Power[b] [mW]		37.92	86.40	169[a]	642	20.48[a]	219[c]	240	549	307[a]
Throughput[b]	Ch. Est. [MEst/s]	N/A	N/A	–	–	–	N/A		28.84	
	QRD [MQRD/s]	–	–	10.64	–	–	N/A		39.60	
	Detection [Mb/s]	N/A	N/A	–	49.60	134		150	454.26	
	Total [Mb/s]	58.47	4	–	–	–	10.8		367.88	
Area. eff.[b]	Ch. est. [KEst/s/KG]	N/A	N/A	–	–	–	N/A		10.45	27.34[a]
	QRD [KQRD/s/KG]	–	–	12.91[a]	–	–	N/A		14.35	37.54[a]
	Detection [Kb/s/KG]	N/A	N/A	–	41.33	1890[a]	N/A		165	431[a]
	Total [Kb/s/KG]	292.34[a]	N/A	–	–	–	1.81	N/A	133	349[a]
Energy[b]	Ch. est. [nJ/Est]	N/A	N/A	15.85[a]	–	–	N/A		12.70	9.58[a]
	QRD [nJ/QRD]	–	–	–	–	–	N/A		15.27	7.96[a]
	Detection [nJ/b]	N/A	N/A	–	1.49	0.3[a]	N/A		0.99	0.62[a]
	Total [nJ/b]	1.2	43.2	–	–	–	N/A	2.23	1.49	0.83[a]

[a] With data buffers excluded
[b] Normalized to 65 nm with 1.2 V core voltage
[c] Only counted relevant parts of the chip

Table 6.16 Comparison of the cell array with programmable platforms

Platform	08 [26]		12 [46]		09 [48]	10 [41]	12 [43]	This work	
	FPGA	DSP	CPU	GPU	GPU	GPU	GPU	Reconfig.	
Antenna	4 × 4		4 × 4		4 × 4	2×2	4 × 4	4 × 4	
Modulation (QAM)	16		64		64	16	64	64	
Mapping (CE|QRD|DT)	–|–|✓		–|–|✓		–|–|✓	–|–|✓	–|✓|✓	✓|✓|✓	
Technology [nm]	130	180	45	40	65	80	40	65	
Area [mm^2]	26[c]	96[c]	296	529	196	N/A	306.82[c]	8.88	
Gate count [KG]	N/A	N/A	1.94e5	7.75e5	1.26e5[c]	2.39e5[c]	4.5e5[c]	2760	1055[a]
Frequency [MHz]	251	200	3070	1150	1900	920	1150	500	
Power[b] [mW]	624[c]	311[c]	101e3	557e3	137e3	9600[c]	323e3[c]	549	307[a]
Throughput[b] — Ch. est. [MEst/s]	–	–	–	–	–	–	–	28.84	
Throughput[b] — QRD [MQRD/s]	–	–	–	–	–	N/A	N/A	39.60	
Throughput[b] — Detection [Mb/s]	163	10.14	0.18	66.09	12.68	44.38	10.58	454.26	
Throughput[b] — Total [Mb/s]	–	–	–	–	–	–	–	367.88	
Area. eff.[b] — Ch. est. [KEst/s/KG]	–	–	–	–	–	–	–	10.45	27.34[a]
Area. eff.[b] — QRD [KQRD/s/KG]	–	–	–	–	–	N/A	N/A	14.35	37.54[a]
Area. eff.[b] — Detection [Kb/s/KG]	N/A	N/A	9.29e-4	0.0853	0.1	0.186[c]	0.0235[c]	165	431[a]
Area. eff.[b] — Total [Kb/s/KG]	–	–	–	–	–	–	–	133	349[a]
Energy[b] — Ch. Est. [nJ/Est]	–	–	–	–	–	–	–	12.70	9.58[a]
Energy[b] — QRD [nJ/QRD]	–	–	–	–	–	N/A	N/A	15.27	7.96[a]
Energy[b] — Detection [nJ/b]	1.32[c]	9.15[c]	3.88e5	1.24e5	2.47e5	2.13e3[c]	3.79e6[c]	0.99	0.62[a]
Energy[b] — Total [nJ/b]	–	–	–	–	–	–	–	1.49	0.83[a]

[a] With data buffers excluded
[b] Normalized to 65 nm with 1.2 V core voltage
[c] Only counted relevant parts of the chip

Table 6.17 Comparison of the cell array with ASIC implementations.

	11 [13]	13 [35]	10 [9]	13 [44]	10 [33]	13 [38]	11 [10]	This work	
Platform	ASIC							Reconfig.	
Antenna	—	—	4 × 4	4 × 4	4 × 4	4 × 4	4 × 4	4 × 4	
Modulation (QAM)	—	—	—	—	64	64	64	64	
Mapping (CE|QRD|DT)	√|—|—	√|—|—	—|√|—	—|√|—	—|—|√	—|—|√	√|√|√	√|√|√	
Technology [nm]	65	65	180	130	130	130	90	65	
Area [mm²]	$0.68^{a,c}$	$2.56^{a,c}$	N/A	0.3^a	3.9^a	N/A	2.02^a	8.88	
Gate count [KG]	$325^{a,c}$	$563^{a,c}$	127.5^a	36^a	491^a	340^a	505^a	2760	1055^a
Frequency [MHz]	250	70	400	278	137.5	417	114	500	
Power^b [mW]	$154^{a,c}$	$73.16^{a,c}$	6.1^a	19.92^a	63.6^a	55^a	59.07^a	549	307^a
Throughput^b — Ch. Est. [MEst/s]	78	11.88	—	—	—	—	N/A	28.84	
Throughput^b — QRD [MQRD/s]	—	—	7.91	13.9	—	—	39.46	39.60	
Throughput^b — Detection [Mb/s]	—	—	—	—	2200	2000	N/A	454.26	
Throughput^b — Total [Mb/s]	—	—	—	—	—	—	947	367.88	
Area. eff.^{b} — Ch. est. [KEst/s/KG]	$240^{a,c}$	$21.09^{a,c}$	—	—	—	—	N/A	10.45	27.34^a
Area. eff.^{b} — QRD [KQRD/s/KG]	—	—	62.06^a	386^a	—	—	78.14^a	14.35	37.54^a
Area. eff.^{b} — Detection [Kb/s/KG]	—	—	—	—	4480^a	5882^a	N/A	165	431^a
Area. eff.^{b} — Total [Kb/s/KG]	—	—	—	—	—	—	1875^a	133	349^a
Energy^b — Ch. est. [nJ/Est]	$1.97^{a,c}$	$6.84^{a,c}$	—	—	—	—	N/A	12.70	9.58^a
Energy^b — QRD [nJ/QRD]	—	—	0.53^a	2.87^a	—	—	N/A	15.27	7.96^a
Energy^b — Detection [nJ/b]	—	—	—	—	0.058^a	0.055^a	N/A	0.99	0.62^a
Energy^b — Total [nJ/b]	—	—	—	—	—	—	2.07^a	1.49	0.83^a

[a] With data buffers excluded

[b] Normalized to 65 nm with 1.2 V core voltage

[c] Scaled up to 4 × 4 MIMO configuration: $\{area, power\} \propto d$, where $d = 4/\#\text{Rx-antenna}$

interesting to note that GPU implementations only achieve a maximum of 66 Mb/s detection throughput, even though they are equipped with enormous parallel computing capacity. For example, the Nvidia Tesla C2070 GPU used in [43] consists of 14 stream multiprocessors, each containing 32 CUDA cores running at 1.15 GHz, and 6 GB of global memory. The low throughput of GPU implementations is mainly caused by the essential difference between wireless baseband processing and graphic computing. GPUs are competent for the latter one. In the wireless communication millions of vectors/matrices with small size have to be handled in parallel, whereas in the graphic computing a large matrix needs to be processed just once in a single application. According to [41, 43, 48], bottlenecks in their design are the limited register resources and memory access bandwidth for each thread processing. In consequence, it is hard to get even half of the CUDA cores utilized for the mapped processing tasks. These show the importance of architectural customization for intended application domains, although algorithm selections and mapping optimization may affect implementation results.

Furthermore, compared to ASIC solutions in Table 6.17, 1.7–13.6 times of difference in area efficiency is observed for each individual task mapping. The result of Löfgren et al. [35] reveals a slightly lower efficiency value than that of the cell array, since it performs data-tone channel estimation in time-domain by reconstructing channel impulse response. Compared to the adopted frequency-domain H interpolation that relies on the correlation properties between neighboring subcarriers, time-domain channel reconstruction leads to better estimation results especially when cyclic prefix is long. However, this performance gain comes at the cost of higher computational complexity and lower throughput, due to involved time-frequency-domain transformations and iterative processing during channel reconstruction.

Energy Efficiency

Besides the area and throughput evaluation, energy consumption per operation is another important measure for baseband processing. In comparison to related implementations in Table 6.15, similar energy figures are observed. However, it should be mentioned that the cell array operates in a more complicated system (4×4 MIMO vs. 2×2 in [12, 28] and 4×2 in [11]) and has more tasks assigned at the same time. Compared to ASICs, the cell array consumes 1.4–15 times more energy for performing each individual task, whereas a 1.3 and 9 times energy gain is obtained in comparison to FPGA and DSP solutions, which support only up to 16-QAM detection. Moreover, its energy efficiency outperforms GPU and CPU approaches by more than 5 orders of magnitude. Such high energy efficiency is achieved mainly by three key hardware developments in the array: architecture partitioning for attaining efficient vector and scalar processing without frequent data alignments, vector enhancements in Tile-0 for reducing register access, and flexible memory access schemes in memory cells for relieving most of the non-computational operations from processing cores.

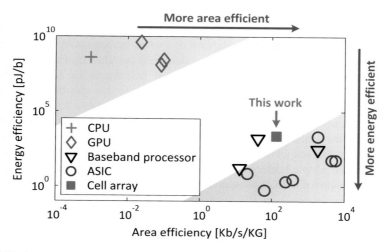

Fig. 6.39 Area and energy efficiency of the cell array in comparison to other hardware platforms

To better visualize the position of the cell array in comparison to other hardware platforms, implementations presented in Tables 6.15, 6.16 and 6.17 are plotted in an area-energy chart, see Fig. 6.39. It clearly shows that the cell array is superior to programmable platforms and achieves an ASIC-like area and energy efficiency. It is worthwhile to re-emphasize here that the presented solution, contrasts to other works, is capable of performing all three MIMO processing tasks in the target 4×4 MIMO 64-QAM 20 MHz LTE-A system.

Hardware Flexibility

In addition to the efficiency analysis, this section discusses the flexibility of the cell array. Among the three architecture categories, programmable platforms offer the greatest flexibility, while ASICs are designed for specific system setups but reveal the highest hardware efficiency. The baseband processors provide palatable flexibility-efficiency trade-offs between the two aforementioned platforms. On one hand, they offer good flexibility as programmable platforms do, but often require more sophisticated software developments, such as low-level programming in domain-specific languages and manual algorithm mappings [29, 40, 47]. On the other hand, their hardware efficiency is largely improved compared to programmable platforms, thanks to their architecture customization and instruction-level acceleration [17]. In this study, the flexibility is demonstrated by time-multiplexing three different tasks onto a reconfigurable cell array. Additionally, by making use of the dynamic resource allocations, such as loading different programs and configurations to processing and memory cells respectively, the platform has the potential to support other system configurations. Examples include processing of different modulation and antenna setups; supporting of different standards; mapping of different algorithms, e.g., non-sorted or iterative-sorted QRD and linear MMSE or SSFE [37] detection; run-time adaption of system performance, e.g., adjusting the frequency of channel estimation and detection parameters. Furthermore, the platform is extensible, thanks to the tile-based heterogeneous and hierarchical resource deployments. For example, larger antenna setups can be supported by extending resource cells and the bandwidth of local links, higher throughput can be achieved by doubling the number of tiles, and system performance can be improved by extending the scalar processing tile (Tile-3) with Log Likelihood Ratio (LLR) unit to perform soft-output data detection [17]. Based on the list of candidate vectors generated in the adopted detection algorithm, a searching unit is needed to find bit-level vectors required in LLR computations. Other scalar operations can be mapped onto the generic processor in Tile-3.

In the following section, the flexibility of the cell array is further illustrated by mapping a hybrid decomposition scheme for performing channel matrix pre-processing. It is aimed to provide a wide range of performance-complexity trade-offs for coping with constantly changing wireless channels. Briefly, the presented scheme dynamically switches between the brute-force SQRD (Sect. 6.2.2) and a low-complexity group-sort QR-update scheme, based on the instantaneous channel condition.

6.6   Adaptive Channel Pre-processor

For the discussions in the previous sections, the propagation channel was assumed to be quasi-static within one LTE-A time slot (0.5 ms). However, Channel state information (CSI) of real-world radio channels are rarely constant because of Doppler induced channel changes and multi-path propagation. Outdated CSI introduces additional interference to the following symbol detection, resulting in drastic degradation of MIMO performance. Thus, frequent CSI update and the corresponding channel matrix pre-processing are highly desirable in wireless communication systems to provide symbol detectors with adequate channel knowledge.

Using the channel's time correlation, tracking of CSI changes can be achieved using low-complexity decision-directed algorithms such as Least Mean Square (LMS), Recursive Least Square (RLS), and Kalman filtering [18, 20]. Nevertheless, continuous CSI tracking has not been widely adopted in practical systems. This is due to the fact that each CSI update requires compute-intensive channel matrix pre-processing, either QRD or channel matrix inversion, which has computational complexity of $\mathcal{O}(N^3)$ for an $N \times N$ MIMO system and consumes more energy than that of symbol detectors (see Table 6.17). To address the complexity and energy issue, [20] proposed an approximated QRD method to avoid exact tone-by-tone QRD computations during successive channel matrix updates. The presented method is based on the assumptions that (1) the orthogonality of column vectors in Q remains unchanged during successive CSI updates; and (2) any change in channel

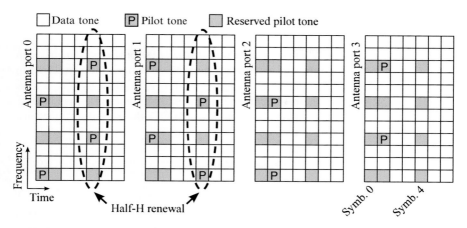

Fig. 6.40 Scattered pilot pattern of LTE-A for four antenna ports. Symbol positions of half-H renewals are *circled in dashed lines*

matrix may be represented as norm value variations in $\boldsymbol{R}$. Although this tracking-$\boldsymbol{R}$ (hold-$\boldsymbol{Q}$) scheme achieves a substantial complexity reduction, it results in a huge performance loss due to the out-of-date $\boldsymbol{Q}$ information, especially in fast-changing channels. Additionally, channel sorting was not considered in [20].

In this section, an adaptive channel matrix pre-processor using a hybrid decomposition scheme with group-sort QR-update strategy [56] is presented and mapped onto the cell array. By fully exploiting the property of the LTE-A pilot pattern, i.e., CSI of only antenna ports 0 and 1 are changed during half-H renewals (Fig. 6.40), the presented QR-update scheme computes exact $\boldsymbol{Q}$ and $\boldsymbol{R}$ matrices using only one Givens rotation. Compared to brute-force QRDs, this update strategy significantly reduces the computational complexity, while preserving the accuracy by avoiding approximations as in the aforementioned tracking algorithms. To obtain the low-complexity benefit of the introduced update scheme in the context of SQRD, an effective group-sort algorithm is introduced for channel reordering. The underlying idea is to restrict the sorting into groups of antenna ports, wherein a two-step (intra- and inter-group) sorting is applied to approximate the optimal detection order. Using the group-sort method, applicability of the QR-update is significantly expanded with negligible performance degradation compared to the precise sorting counterpart.

To ease the discussion, the SQRD algorithm is used as a case study in the following sections. However, the presented methods can be applied to the MMSE-SQRD case by working with the augmented matrix $\underline{\hat{\boldsymbol{H}}}$ instead of $\hat{\boldsymbol{H}}$.

6.6.1   QR-Update Scheme

If only parts of the matrix columns alter over time, QRD of the new matrix can be performed in a more efficient way than a brute-force computation (referred to as Case-I), i.e., starting from scratch using Algorithm 1. Inspired by this, a low-complexity QR-update scheme is adopted during half-H renewals. Specifically, the presented scheme starts with the brute-force SQRD during full-H renewals, expressed with a subscript "old" as

$$\hat{H}_{p,\text{old}} = Q_{\text{old}} R_{\text{old}}, \tag{6.27}$$

where $\hat{H}_p$ is the permuted channel matrix (6.5). During half-H renewals, $\hat{H}_{p,\text{new}}$ is obtained by updating two columns of $\hat{H}_{p,\text{old}}$. Although orthogonal vectors in Q_{old} may no longer triangularize $\hat{H}_{p,\text{new}}$, it may still have vectors pointing in the correct directions. As a consequence, the new R matrix, denoted as $\tilde{R}_{\text{new}}$, can be expressed using $\hat{H}_{p,\text{new}}$ and Q_{old} as

$$\tilde{R}_{\text{new}} = Q_{\text{old}}^H \hat{H}_{p,\text{new}}. \tag{6.28}$$

Due to the outdated Q_{old}, $\tilde{R}_{\text{new}}$ is no longer an upper-triangular matrix but may still reveal some upper-triangular properties depending on the positions of the two renewed columns. Specifically, if column changes take place at the right-most of $\hat{H}_{p,\text{new}}$, only one element in the lower triangular part of $\tilde{R}_{\text{new}}$ (i.e., $\tilde{r}_{\text{new}(4,3)}$ for 4×4 MIMO) becomes non-zero. This implies that triangularization of $\tilde{R}_{\text{new}}$ can be significantly simplified by nulling the single non-zero element rather than operating on all columns afresh. Defining G as a complex-valued nulling matrix

$$G = \begin{bmatrix} 1 & 0 & 0 & 0 \\ 0 & 1 & 0 & 0 \\ 0 & 0 & c & s^* \\ 0 & 0 & -s^* & c^* \end{bmatrix}, \tag{6.29}$$

$\tilde{R}_{\text{new}}$ can be triangularized by computing $G\tilde{R}_{\text{new}}$. In (6.29), $(\cdot)^*$ denotes the complex conjugate. c and s are defined as

$$c = \tilde{r}_{\text{new}(3,3)}^* / z,$$
$$s = \tilde{r}_{\text{new}(4,3)}^* / z, \tag{6.30}$$
$$z = \left(|\tilde{r}_{\text{new}(3,3)}|^2 + |\tilde{r}_{\text{new}(4,3)}|^2 \right)^{1/2}.$$

This nulling process is commonly referred to as Givens rotation [19]. After triangularizing $\tilde{R}_{\text{new}}$, exact Q_{new} and R_{new} are obtained, expressed as

$$Q_{\text{new}} = \left(GQ_{\text{old}}^H\right)^H,$$

$$R_{\text{new}} = G\tilde{R}_{\text{new}} = G\left(Q_{\text{old}}^H \hat{H}_{p,\text{new}}\right). \tag{6.31}$$

It should be pointed out that the lower-right diagonal element of R_{new} in (6.31), i.e., $r_{\text{new}(4,4)}$, has been transformed from real to complex-valued domain during the QR updates. This can be easily resolved by performing an additional real-valued domain-transformation using another nulling matrix G' [36], if real-valued diagonal elements are required. The matrix G' is defined as

$$G' = \begin{bmatrix} 1 & 0 & 0 & 0 \\ 0 & 1 & 0 & 0 \\ 0 & 0 & 1 & 0 \\ 0 & 0 & 0 & c' \end{bmatrix}, \tag{6.32}$$

where

$$c' = \tilde{r}_{\text{new}(4,4)}^* / |\tilde{r}_{\text{new}(4,4)}|. \tag{6.33}$$

By combining the traditional brute-force approach and the QR-update scheme, a hybrid decomposition algorithm is formed. Depending on the run-time condition of the channel reordering, the two schemes can be dynamically switched to reduce the computational complexity. Obviously, the complexity reduction depends on the applicability of the QR-update. Intuitively, the position of antenna ports 0 and 1 can be fixed to the right-most part of $\hat{H}_{p,\text{new}}$ in order to obtain a maximum complexity gain, since it completely avoids brute-force computation during half-H renewals. However, the advantage of channel reordering, for improving detection performance, is lost. This approach is referred to as Case-II. On the other hand (Case-III), the applicability of the QR-update is dramatically reduced if channel columns are permuted based on the optimal detection order without considering the position of renewed channel columns. For example, considering the 4×4 MIMO LTE-A, only $(2!2!)/4! = 1/6$ of sorting combinations meet the required update condition, thus limiting the complexity reduction. As a consequence, a smart scheduling strategy is needed to explore the low-complexity potential of the QR-update, while still retaining the performance gain of the optimal channel reordering.

6.6.2   Group-Sort Algorithm

To fulfill the aforementioned requirement, an effective group-sort algorithm is introduced for channel reordering. Instead of operating on individual columns, sorting of $\hat{H}$ is applied on two virtual groups, wherein columns associated with antenna ports 0 and 1 are tied together. This way, the number of combinations of

Table 6.18 Case-I–IV of the presented hybrid decomposition algorithm

	Channel reordering	Brute-force [%]	QR-update [%]
Case-I	Optimal ordering with precise sorting	100	0
Case-II	Fixed order for antenna ports 0 and 1	0	100
Case-III	Optimal ordering with precise sorting	83.33	16.67
Case-IV	Group-sort	50	50

"columns" is reduced from 4! to 2!. Consequently, the probability of having both altered columns at the right-most part of $\hat{\boldsymbol{H}}_{p,\mathrm{new}}$ is increased by 3 times, from $1/6$ to $1/2$. To reduce errors caused by sub-optimal sorting sequences, a two-step sorting scheme is adopted. First, the sorting between groups is based on the total energy of bundled columns as

$$\mathcal{I} = \arg\max_{i=\{0,1\},\{2,3\}} \sum_i \|\hat{\boldsymbol{h}}_{p(i)}\|^2, \tag{6.34}$$

where $\mathcal{I}$ contains inter-sorted group indexes, e.g., $\mathcal{I} = \{0, 1\}$ if antenna ports 0 and 1 correspond to the strongest channels. Second, the two columns within each group, e.g., indexes within $\mathcal{I}$, are intra-sorted based on the energy of individual columns. To summarize, Table 6.18 lists applicability of all four cases of the hybrid decomposition algorithm. The presented group-sort method is denoted by Case-IV.

6.6.3   Algorithm Evaluation and Operation Analysis

To illustrate the effectiveness of the presented algorithm, the same simulation setup as the one presented in Sect. 6.3.1 with parameters in Table 6.1 is employed, except that the variant channel modeling is used in this case for emulating time-variations of radio channels. At 2.6 GHz carrier frequency, the maximum Doppler frequency of 70 Hz corresponds to a speed of 29 km/h. Channel models with higher mobility, such as ETU model with 300 Hz Doppler frequency, are not considered in this study, as smaller MIMO configurations (e.g., 2×2) or lower modulation schemes (e.g., QPSK) are expected to be used to mitigate serious interference induced by fast channel variations. To minimize performance influences from channel estimation and symbol detection, perfect channel knowledge is assumed at the receiver and the near-ML FSD algorithm is used for symbol detection.

Performance of the presented group-sort QR-update and aforementioned cases are shown in Fig. 6.41. Note that Case-III has the same performance as the brute-force approach and is used as a reference for FER measurements. The difference in performance is remarkable between Case-III and the case where no QRDs are performed during half-H renewals (the upper curve in Fig. 6.41), indicating the importance of performing CSI and QR updates even for channels with moderate Doppler shifts. Additionally, adoption of channel reordering during QR decom-

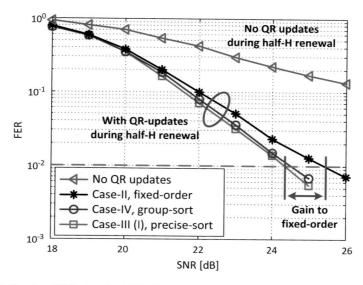

Fig. 6.41 Simulated FERs in a 4×4 MIMO LTE-A downlink using 3GPP EVA-70 channel model with 64-QAM modulation

Table 6.19 Complexity of computations in the hybrid decomposition algorithm

Complexity	Computation	Multiplication	$1/\sqrt{x}$
C_1	QRD (6.27)	$N^3 + 2N^2$	N
C_2	$\boldsymbol{Q}_{\text{old}}^H \hat{\boldsymbol{H}}_{p,\text{new}}$ (6.28)	$\frac{1}{2}N^3$	0
C_3	Triangularization (6.29)–(6.31)	$N^2 + 2N$	1
C_4	Sorting (Algorithm 1 lines 3 and 12)	$6N$	0
C_5	Sorting (6.34)	$4N$	0

position improves performance to that of the fixed-order approach, e.g., 1.1 dB difference between Case-II and III at FER $= 10^{-2}$. Furthermore, the group-sort approach has only small performance degradation of about 0.2 dB compared to Case-III, however, with a large complexity reduction as analyzed in the following.

Table 6.19 summaries complexity (C) of computations (6.27)–(6.31) and (6.34) for an $N \times N$ MIMO system. To perform the brute-force decomposition (6.27), MGS algorithm (Algorithm 1) is considered which has a complexity of C_1. Computations required for both (6.28) and (6.31) have a total complexity of $C_2 + C_3$, which is significantly lower than C_1, e.g., by about 42 % for $N = 4$. Note that the product of $\boldsymbol{Q}_{\text{old}}^H \hat{\boldsymbol{H}}_{p,\text{new}}$ in (6.28) requires only half of the matrix computations during QR updates, since only two columns change in $\hat{\boldsymbol{H}}_{p,\text{new}}$. The complexity of the precise-iterative-sort and the group-sort approach is denoted as C_4 and C_5, respectively. Based on this analysis and in reference to Case-I, Table 6.20 shows the complexity reduction versus performance degradation of Case-II–IV for a 4×4 MIMO system. It shows that Case-II reduces the computational complexity by 53 %. Additionally,

Table 6.20 Complexity and performance comparisons of Case-I–IV

	Complexity	Complexity reduction	SNR degradation
Case-I	$C_1 + C_4$	− (ref.)	− (ref.)
Case-II	$C_2 + C_3$	53 %	1.1 dB
Case-III	$\frac{5}{6}C_1 + \frac{1}{6}(C_2 + C_3) + C_4$	6 %	0 dB
Case-IV	$\frac{1}{2}C_1 + \frac{1}{2}(C_2 + C_3) + C_5$	23 %	0.2 dB

combining the group-sort and the QR-update scheme results in more palatable trade-offs, i.e., 23 % complexity reduction for only 0.2 dB performance degradation.

To further evaluate hardware friendliness of the presented algorithm and the possibility of being mapped onto the cell array, operations required in the four computations (Table 6.19) are profiled, see Table 6.21. It clearly shows that all operations required in the group-sort QR-update algorithm are shared with that of the brute-force method. This implies that extensive hardware reuse is possible. Additionally, over 95 % of the operations are at vector-level, representing a high degree of DLP that can be exploited to achieve high processing throughput.

6.6.4   Implementation Results and Discussion

It is straightforward to implement the adaptive channel matrix pre-processor onto the cell array using the hybrid decomposition scheme, Case-I–IV in Table 6.18, since all the required operations (Table 6.21) have already been mapped for the SQRD computation. Similar to the task mapping scheme presented in the previous section, multi-subcarrier processing is adopted for attaining high throughput and is scheduled based on the LTE-A resource block. Worth mentioning is that Givens rotation (6.31) can be implemented in different ways, such as using conventional arithmetic or through a series of CORDIC operations [36]. In Chap. 5, the CORDIC algorithm is mapped onto the scalar processing cell for computing the phase value of received symbols. However, each computation requires several clock cycles to complete because of the iterative nature of the CORDIC. Evidently, adopting the CORDIC approach in the QR-update computation would make the triangularization process an implementation bottleneck. Hence, Givens rotation is realized by using conventional arithmetic, where c and s in (6.30) are computed by the dedicated inverse square root unit available in Tile-3.

Table 6.22 summaries implementation results for the brute-force and the QR-update computations. Operating at 500 MHz, processing throughput of the QR-update is 2.6 times higher than that of the brute-force approach. Moreover, it reduces the energy consumption by 1.9 times. It should be pointed out that further energy reduction is possible if fine-grained low power design techniques are employed. For example, the computation of Q_{new} (6.31) can be performed more efficiently if half of the SIMD core in Tile-0 could be clock gated, because of the zero elements in the nulling matrix G.

Table 6.21   Operation profile for $N = 4$

	Vector operations			
	$A \cdot B$	$A \odot B$	$A \pm B$	$1/\sqrt{x}$
QRD (6.27)	17	4	6	4
$Q_{\text{old}}^H \hat{H}_{p,\text{new}}$ (6.28)	8	0	0	0
Triangularization (6.29)–(6.31)	10	0	0	1
Sorting (6.34)	4	0	0	0

Table 6.22   Performance summary

	Clock Cycle/Op	Throughput [MQRD/s]	Energy[a] [nJ/QRD]
Brute-force SQRD	12.63	39.60	7.96
Group-sort QR-update	4.83	103.45	4.29

[a] With data buffers excluded

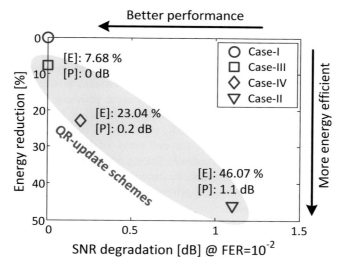

Fig. 6.42 Energy (E) and performance (P) trade-off for Case-I–IV of the hybrid decomposition scheme

Figure 6.42 presents design trade-offs between energy and performance for Case-I–IV of the hybrid decomposition scheme. Taking the brute-force QRD (Case-I) as a reference, numbers on the horizontal axis measures the SNR degradation for reaching the target 10^{-2} FER, while the percentage of energy reduction is shown on the vertical axis. Accordingly, algorithms having their coordinates towards the bottom-left corner are desired. Figure 6.42 clearly shows that the presented group-sort QR-update scheme (Case-IV) achieves a good compromise, i.e., trading 0.2 dB performance for 23 % energy reduction. In the case of energy-constrained systems, the fixed-order scheme (Case-II) can be adopted to further reduce the energy consumption, i.e., by 46 % in total, whereas the precise-sort scheme (Case-III) can be used if high performance is demanded. Benefiting from the flexibility of the cell array, the selection of the algorithm can be made at system run-time, depending on instantaneous channel condition, performance requirement, and power budget.

6.7 Summary

This chapter presents a reconfigurable baseband processor designed based on the heterogeneous cell array architecture. The performance and flexibility of the presented solution is exhibited by mapping three crucial baseband processing blocks onto the processor, while the capability of real-time processing in an LTE-A downlink is demonstrated. Such high processing performance is achieved by algorithm–architecture co-design. On the algorithm side, more than 98 % of total operations in all three tasks are unified to a vector-level, enabling extensive parallel processing and resource sharing for attaining high hardware efficiency. Further achievements in area and energy efficiency are enabled by architectural developments, including the heterogeneous resource deployments, vector enhancements of the processing core, and flexible self-governed memory data access schemes. Implementation results show that the presented processor bridges the gap between conventional platforms. The processor provides enormous design flexibility and scalability like programmable platforms, while approaching the area and energy efficiency of task specific ASIC solutions. In addition to the multi-task MIMO processing, the flexibility of the cell array is demonstrated by mapping an adaptive channel matrix pre-processor. Taking advantage of dynamic resource allocations, a wide range of performance-complexity trade-offs are provided, so that an appropriate pre-processing algorithm can be selected at run-time based on instantaneous channel condition.

References

1. 3GPP TS 36.101 V11.4.0: user equipment (UE) radio transmission and reception (Release 11), Mar 2013. http://www.3gpp.org/ftp/Specs/archive/36_series/36.101/36101-b40.zip
2. 3GPP TS 36.212 V11.2.0: multiplexing and channel coding (Release 11), Feb 2013. http://www.3gpp.org/ftp/Specs/archive/36_series/36.212/36212-b20.zip
3. L. Bahl, J. Cocke, F. Jelinek, J. Raviv, Optimal decoding of linear codes for minimizing symbol error rate. IEEE Trans. Inf. Theory **20**(2), 284–287 (1974)
4. L.G. Barbero, J.S. Thompson, Fixing the complexity of the sphere decoder for MIMO detection. IEEE Trans. Wirel. Commun. **7**(6), 2131–2142 (2008)
5. C. Bernard, F. Clermidy, A low-power VLIW processor for 3GPP-LTE complex numbers processing, in *Design, Automation Test in Europe Conference Exhibition (DATE)*, Mar 2011, pp. 1–6
6. C. Berrou, A. Glavieux, Near optimum error correcting coding and decoding: turbo-codes. IEEE Trans. Commun. **44**(10), 1261–1271 (1996)

7. A. Burg, et al., VLSI implementation of MIMO detection using the sphere decoding algorithm. IEEE J. Solid State Circuits **40**(7), 1566–1577 (2005)
8. J. Byrne, Tensilica DSP targets LTE advanced, Mar 2011. http://www.tensilica.com/uploads/pdf/MPR_BBE64.pdf
9. R.C.H. Chang, C.H. Lin, K.H. Lin, C.L. Huang, F.C. Chen, Iterative QR decomposition architecture using the modified Gram-Schmidt algorithm for MIMO systems. IEEE Trans. Circuits Syst. Regul. Pap. **57**(5), 1095–1102 (2010)
10. P.L. Chiu, L.Z. Huang, L.W. Chai, C.F. Liao, Y.H. Huang, A 684Mbps 57mW Joint QR decomposition and MIMO processor for 4×4 MIMO-OFDM systems, in *IEEE Asian Solid State Circuits Conference (ASSCC)*, Nov 2011, pp. 309–312
11. F. Clermidy, et al., A 477mW NoC-based digital baseband for MIMO 4G SDR, in *IEEE International Solid-State Circuits Conference (ISSCC)*, Feb 2010, pp. 278–279
12. V. Derudder, et al., A 200Mbps+ 2.14nJ/b digital baseband multi processor system-on-chip for SDRs, in *IEEE Symposium on VLSI Circuits (VLSIC)*, 2009, pp. 292–293
13. I. Diaz, B. Sathyanarayanan, A. Malek, F. Foroughi, J.N. Rodrigues, Highly scalable implementation of a robust MMSE channel estimator for OFDM multi-standard environment, in *IEEE Workshop on Signal Processing Systems (SiPS)*, 2011, pp. 311–315
14. O. Edfors, M. Sandell, J.J. van de Beek, S.K. Wilson, P.O. Börjesson, OFDM channel estimation by singular value decomposition. IEEE Trans. Commun. **46**(7), 931–939 (1998)
15. F. Edman, V. Öwall, A scalable pipelined complex valued matrix inversion architecture. in *IEEE International Symposium on Circuits and Systems (ISCAS)*, vol. 5, 2005, pp. 4489–4492
16. M.D. Ercegovac, L. Imbert, D.W. Matula, J.M. Muller, G. Wei, Improving Goldschmidt division, square root, and square root reciprocal. IEEE Trans. Comput. **49**(7), 759–763 (2000)
17. R. Fasthuber, et al., Exploration of soft-output MIMO detector implementations on Massive parallel processors. J. Signal Process. Syst. **64**, 75–92 (2011) J. Signal Process. Syst. **64**, 75–92 (2011)
18. S. Gifford, C. Bergstrom, S. Chuprun, Adaptive and linear prediction channel tracking algorithms for mobile OFDM-MIMO applications, in *IEEE Military Communications Conference (MILCOM)*, vol. 2, Oct 2005, pp. 1298–1302
19. G.H. Golub, C.F. Van Loan, *Matrix Computations*, 3rd edn. (Johns Hopkins University Press, Baltimore, Maryland, 1996)
20. L. Gor, M. Faulkner, Power reduction through upper triangular matrix tracking in QR detection MIMO receivers, in *IEEE 64th Vehicular Technology Conference (VTC)*, Sept 2006, pp. 1–5
21. S. Granlund, L. Liu, C. Zhang, V. Öwall, A low-latency high-throughput soft-output signal detector for spatial multiplexing MIMO systems. Microprocess. Microsyst. 2015. http://www.sciencedirect.com/science/article/pii/S0141933115000034
22. Z. Guo, P. Nilsson, Algorithm and implementation of the K-best sphere decoding for MIMO detection. IEEE J. Sel. Areas Commun. **24**(3), 491–503 (2006)
23. S. Haene, D. Perels, A. Burg, A real-time 4-Stream MIMO-OFDM transceiver: system design, FPGA implementation, and characterization. IEEE J. Sel. Areas Commun. **26**(6), 877–889 (2008)
24. R.W. Heath, A. Paulraj, Antenna selection for spatial multiplexing systems based on minimum error rate, in *IEEE International Conference on Communications (ICC)*, vol. 7, 2001, pp. 2276–2280
25. M.H. Hsieh, C.H. Wei, Channel estimation for OFDM systems based on comb-type pilot arrangement in frequency selective fading channels. IEEE Trans. Consum. Electron. **44**(1), 217–225 (1998)
26. X. Huang, C. Liang, J. Ma, System architecture and implementation of MIMO sphere decoders on FPGA. IEEE Trans. Very Large Scale Integr. VLSI Syst. **16**(2), 188–197 (2008)
27. Z.-Y. Huang, P.-Y. Tsai, Efficient implementation of QR decomposition for gigabit MIMO-OFDM systems. IEEE Trans. Circuits Syst. Regul. Pap. **58**(10), 2531–2542 (2011)
28. J. Janhunen, O. Silven, M. Juntti, M. Myllyla, Software defined radio implementation of K-best list sphere detector algorithm, in *International Conference on Embedded Computer Systems: Architectures, Modeling, and Simulation (SAMOS)*, July 2008, pp. 100–107

29. J. Janhunen, T. Pitkanen, O. Silven, M. Juntti, Fixed- and floating-point processor comparison for MIMO-OFDM detector. IEEE J. Sel. Top. Sign. Proces. **5**(8), 1588–1598 (2011)
30. Y. Kim, R.N. Mahapatra, I. Park, K. Choi, Low power reconfiguration technique for coarse-grained reconfigurable architecture. IEEE Trans. Very Large Scale Integr. VLSI Syst. **17**(5), 593–603 (2009)
31. C. Kozyrakis, D. Patterson, Vector vs. superscalar and VLIW architectures for embedded multimedia benchmarks, in *35th Annual IEEE/ACM International Symposium on Microarchitecture*, 2002, pp. 283–293
32. H. Lee, C. Chakrabarti, T. Mudge, A low-power DSP for wireless communications. IEEE Trans. Very Large Scale Integr. VLSI Syst. **18**(9), 1310–1322 (2010)
33. L. Liu, F. Ye, X. Ma, T. Zhang, J. Ren, A 1.1-Gb/s 115-pJ/bit configurable MIMO detector using 0.13-μCMOS technology. IEEE Trans. Circuits Syst. Express Briefs **57**(9), 701–705 (2010)
34. L. Liu, J. Löfgren, P. Nilsson, Area-efficient configurable high-throughput signal detector supporting multiple MIMO modes. IEEE Trans. Circuits Syst. Regul. Pap. **59**(9), 2085–2096 (2012)
35. J. Löfgren, L. Liu, O. Edfors, P. Nilsson, Improved matching-pursuit implementation for LTE channel estimation. IEEE Trans. Circuits Syst. Regul. Pap. **61**(1), 226–237 (2014)
36. P. Luethi, A. Burg, S. Haene, D. Perels, N. Felber, W. Fichtner, VLSI implementation of a high-speed iterative sorted MMSE QR decomposition, in *IEEE International Symposium on Circuits and Systems (ISCAS)*, 2007, pp. 1421–1424
37. M. Li, et al., Optimizing near-ML MIMO detector for SDR baseband on parallel programmable architectures, in *Design, Automation and Test in Europe (DATE)*, Mar 2008, pp. 444–449
38. M. Mahdavi, M. Shabany, Novel MIMO detection algorithm for high-order constellations in the complex domain. IEEE Trans. Very Large Scale Integr. VLSI Syst. **21**(5), 834–847 (2013)
39. K. Mohammed, B. Daneshrad, A MIMO decoder accelerator for next generation wireless communications. IEEE Trans. Very Large Scale Integr. VLSI Syst. **18**(11), 1544–1555 (2010)
40. A. Nilsson, E. Tell, D. Liu, An 11 mm^2, 70 mW fully programmable baseband processor for mobile WiMAX and DVB-T/H in 0.12μm CMOS. IEEE J. Solid-State Circuits **44**(1), 90–97 (2009)
41. T. Nylanden, J. Janhunen, O. Silven, M. Juntti, A GPU implementation for two MIMO-OFDM detectors, in *International Conference on Embedded Computer Systems: Architectures, Modeling, and Simulation (SAMOS)*, July 2010, pp. 293–300
42. J.M. Rabaey, A. Chandrakasan, B. Nikolic, *Digital Integrated Circuits - A Design Perspective*, 2nd edn. (Prentice Hall, Englewood Cliffs, 2002)
43. S. Roger, C. Ramiro, A. Gonzalez, V. Almenar, A.M. Vidal, Fully parallel GPU implementation of a fixed-complexity soft-output MIMO detector. IEEE Trans. Veh. Technol. **61**(8), 3796–3800 (2012)
44. M. Shabany, D. Patel, P.G. Gulak, A low-latency low-power QR-decomposition ASIC implementation in 0.13 μm CMOS. IEEE Trans. Circuits Syst. Regul. Pap. **60**(2), 327–340 (2013)
45. M. Šimko, D. Wu, C. Mehlfüehrer, J. Eilert, D. Liu, Implementation aspects of channel estimation for 3GPP LTE terminals, in *11th European Wireless Conference*, Apr 2011, pp. 1–5
46. D. Sui, Y. Li, J. Wang, P. Wang, B. Zhou, High throughput MIMO-OFDM detection with graphics processing units, in *IEEE International Conference on Computer Science and Automation Engineering (CSAE)*, vol. 2, May 2012, pp. 176–179
47. M. Thuresson, et al., FlexCore: utilizing exposed datapath control for efficient computing. J. Signal Process. Syst. **57**(1), 5–19 (2009)
48. M. Wu, S. Gupta, Y. Sun, J.R. Cavallaro, A GPU implementation of a real-time MIMO detector, in *IEEE Workshop on Signal Processing Systems (SiPS)*, Oct 2009, pp. 303–308
49. D. Wübben, J. Rinas, R. Böhnke, V. Kühn, K.D. Kammeyer, Efficient algorithm for detecting layered space-time codes, in *4th International ITG Conference on Source and Channel Coding (SCC)*, Jan 2002, pp. 399–405

50. D. Wübben, R. Böhnke, V. Kühn, K.D. Kammeyer, MMSE extension of V-BLAST based on sorted QR decomposition, in *IEEE 58th Vehicular Technology Conference (VTC)*, vol. 1, 2003, pp. 508–512
51. Y. Xie, W. Wolf, H. Lekatsas, Code compression for embedded VLIW processors using variable-to-fixed coding. IEEE Trans. Very Large Scale Integr. VLSI Syst. **14**(5), 525–536 (2006)
52. C. Yang, D. Marković, A flexible DSP architecture for MIMO sphere decoding. IEEE Trans. Circuits Syst. Regul. Pap. **56**(10), 2301–2314 (2009)
53. S. Ye, S. H. Wong, C. Worrall, Enhanced physical downlink control channel in LTE advanced release 11. IEEE Commun. Mag. **51**(2), 82–89 (2013)
54. C. Zhang, T. Lenart, H. Svensson, V. Öwall, Design of coarse-grained dynamically reconfigurable architecture for DSP applications, in *International Conference on Reconfigurable Computing and FPGAs (ReConFig)*, Dec 2009, pp. 338–343
55. C. Zhang, L. Liu, D. Marković, V. Öwall, A heterogeneous reconfigurable cell array for MIMO signal processing. IEEE Transactions on Circuits and Systems I: Regular Papers, **62**(3), 733–742 (2015)
56. C. Zhang, H. Prabhu, Y. Liu, L. Liu, O. Edfors, V. Öwall, Energy efficient group-sort QRD processor with on-line update for MIMO channel pre-processing. IEEE Trans. Circuits Syst. Regul. Pap. **62**(5), 1220–1229 (2015)

Chapter 7
Future Multi-User MIMO Systems:
A Discussion

Wireless communication technology is evolving at a fast pace to meet requirements of emerging applications, such as ultra-high resolution video, cloud computing, internet of things, etc. For example, it took only 2 years for cellular systems to evolve from LTE to LTE-A, delivering a $10\times$ increase in data rates. Almost at the same time that the first LTE-A service was launched, people are talking about next-generation (5G) wireless communication systems [3]. The coming 5G communication aims to connect tens of billions of devices with some reaching several gigabit-per-second data rates and milliseconds service latency. On the other hand, bandwidth is a scare resource, demanding revolution in wireless communication technologies to achieve these aggressive targets. Technologies being discussed include small-cell networks [7], millimeter wave communication [14], interference cancellation (e.g., full-duplex transmission [2]), advanced wave-forms [6] (e.g., Generalized Frequency Division Multiplexing, Universal Filtered Multi-carrier, Filter-Bank based Multi-Carrier, Bi-orthogonal Frequency Division Multiplexing), Massive MIMO [4, 16], etc. Among those, Massive MIMO has been widely accepted, both in academia and industry, as one of the promising candidates for 5G. 3GPP is developing 3D channel models for this new MIMO technique. Studies for the Time-division duplexing (TDD) Massive MIMO have been initiated for 3GPP Release 13. In this chapter, we will focus on the Massive MIMO technology, discussing its basic concepts, state-of-the-art research progress, key signal processing in the digital baseband, as well as new challenges for designing reconfigurable architectures for its baseband implementation.

© Springer International Publishing Switzerland 2016
C. Zhang et al., *Heterogeneous Reconfigurable Processors for Real-Time Baseband Processing*, DOI 10.1007/978-3-319-24004-6_7

7.1   MIMO Goes to Massive

7.1.1   Massive MIMO Basics

In 5G era, multi-user support is crucial since the number of linked devices will grow explosively. Currently, this has been achieved by multiplexing users in resource domains such as time, frequency, and code. However, the contradiction between the growth in number of mobile devices and ever more precious spectrum poses a fundamental limit. Traditional multi-antenna technology exploits the spatial domain, opening an opportunity to support more users. Unfortunately, the physical size and energy source limitations on portable devices prevent from adding significantly more antennas. To make a clean break with traditional MIMO, a new multi-user multi-antenna scheme, commonly referred to as Massive MIMO (also known as large-scale MIMO, full-dimension MIMO [8]), has been introduced.

As illustrated in Fig. 7.1, the concept of Massive MIMO includes the deployment of antenna array at the base station, having much more elements than in systems being built today, say 100 or more. The base station simultaneously serves a relatively smaller number of User Equipment (UE), e.g., $K \ll M$ in Fig. 7.1. The physically large array with many antenna elements provides huge degree-of-freedom in spatial domain and has some special channel properties that may not be observed in "traditional" multi-antenna systems:

- The large antenna array has large Rayleigh distance, making it possible to focus transmission energy into smaller space to multiplex more users in the same time-frequency resource. For instance, it can separate users not only in direction but also in "depth."
- Large-scale fading can be experienced across the large antenna array, providing diversity in radio signal propagation.

Fig. 7.1 Illustration of a Massive MIMO system with M base station antennas and K single-antenna user equipment

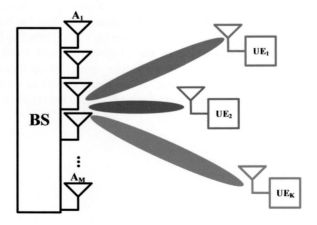

- Under favorable propagation environments, channels between base station and different users become orthogonal as the number of base station antenna elements grows.
- Some channel (or equivalent channel) variations, like small-scale fading, can be averaged out even with simple processing schemes.

As a result, Massive MIMO (with multiple users being served simultaneously in the same time-frequency resource) is capable of delivering orders of magnitude increase in both spectral efficiency and radiated energy efficiency. Furthermore, Massive MIMO has the potential to relax the design requirement for analog front-end circuits by allowing extensive use of low-cost components. For instance the combination of large number of signal paths averages out non-ideal phenomenon in analog components, e.g., phase noise and fabrication variations. Another example of using low-cost analog components is the deployment of hundreds of power amplifiers with output power in the milli-Watt range, instead of those expensive ultra-linear power amplifiers (generally in tens-of-Watt range) used in current systems [4]. Another benefit of multi-user Massive MIMO systems is the reduced latency, since it inherently reduces deep fading that may require complicated retransmission. The scheduling of multiple users can be simplified since the multi-user multiplexing can be performed at the spatial domain only. From the baseband processing perspective, the computational complexity can also be moderated, to a certain extent. Because of the huge degree-of-freedom in service antennas, linear MIMO pre-coding and detection are able to achieve near-optimal performance. Finally, UE are significantly simplified since all the computational-demanding processing will be performed at the base station side.

7.1.2   From Theory to Practice

Because of its promising improvement in system capacity, Massive MIMO has attracted lots of research activities. In the following, we give a brief overview of Massive MIMO related research activities, both in theoretical exploration and practical implementation. The purpose of this overview is to identify critical research problems for the new Massive MIMO systems and to introduce the corresponding digital signal processing techniques. These are important aspects and are the prerequisite for designing high-performance processor architectures for baseband processing.

Theoretical Research

In the level of theoretical analysis and development, substantial research progress has been achieved and reported. Interesting and important topics include performance analysis, spectral and energy efficiency optimization, and resource allocation.

In [10], the uplink performance of a single-cell Massive MIMO system (with M base station antennas) has been analyzed. Utilizing the array and spatial-diversity gain, the transmitted power of each user can be reduced proportional to $\frac{1}{M}$ without obvious performance degradation. The information-theoretical analysis reported that Massive MIMO is able to provide 100 times improvement in power efficiency and ten times in spectrum efficiency [9].

To harvest this promising performance enhancement in real-life prorogation environment, channel feature and modeling in the context of Massive MIMO transmission has to be studied. In [1], channel measurement has been conducted for a 128-antenna Massive MIMO system using both large linear array and a compact cylindrical array. Equipped with low-complexity linear ZF pre-coding, close-to-optimal performance (i.e., performance with i.i.d. channel) can be approached in measured channels.

Another important research aspect is key signal processing algorithms in Massive MIMO systems, including synchronization and calibration, channel estimation, antenna selection, uplink MIMO detection, and downlink beam-forming pre-coding. Currently, the most popular Massive MIMO transmission is based on TDD, where the channel reciprocity is utilized to construct downlink beam-forming with uplink channel estimation. In this case, accurate CSI acquisition becomes crucial to system performance. Other critical issues to be tackled to achieve good overall system performance is the reduction of inter-cell interference, e.g., pilot contamination problems [4]. On the other side, the main research objective in the area of MIMO pre-coding and detection has been approaching optimal performance with low computational complexity [11, 18]. In addition, how to relieve the design requirements for analog components using digital signal processing technology has also received many attentions, e.g., the low Peak-to-Average Power Ratio (PAPR) or constant-envelope pre-coding [13].

Implementation-Related Research

On the other hand, research dedicated to the hardware implementation perspective is still in its infancy. Currently published implementations have been focused on key signal processing blocks in Massive MIMO baseband processing. For instance, researchers at Rice University and Cornell University are working together with the focus on uplink multi-user signal detectors. In [19], they reported an implementation of a Neumann Series expansion-based MMSE detector for a 128-antenna 8-UE system. The achieved detection throughput is 600 Mb/s. Researches at Lund University are investigating efficient implementation of beam-forming pre-coders. The feature of Massive MIMO, i.e., the Gramian of the channel matrix is diagonally dominated, has been explored to develop low-complexity large-size matrix inversion using a Neumann Series approximation [11]. Moreover, the performance-complexity optimization has been performed to find the trade-off between the iteration number and the pre-condition matrix in Neumann Series algorithm. The pre-coder is evaluated using a 65 nm CMOS technology and consumes 125 K gates. Running

Table 7.1 Reported Massive MIMO test-beds

Test-bed	Operating band (GHz)	No. BS antennas	No. users	Hardware platform
LuMaMi	1.2–6	100 (planar array)	10	NI USRP with Xilinx FPGA
Argos	2.4	64 (planar array)	15	WARP v3 Node with Xilinx FPGA
Samsung	1–28	64 (planar array)	N/A	N/A
Ngara	0.8	32 (circular array)	18	APU with Xilinx FPGA

at 420 MHz clock frequency, it takes only 4 μs to finish the pre-coding for a 100-antenna 16-UE system. This low latency is important, in addition to the traditional throughput requirement, for a TDD Massive MIMO system. In [12, 13], Lund researchers reported pre-coders based on antenna reservation and constant envelop techniques that lead to low PAPRs for transmitting signals, while being efficient in terms of complexity, enabling the use of low-cost power amplifiers.

To proof the concept of new wireless communication techniques, it is important to build up test-beds to conduct verification with over-the-air transmission. For Massive MIMO it is even more important, since it heavily relies on propagation environment. However, building up a Massive MIMO test-bed is a challenging task, which consumes lots of resources and requires experienced engineering in both wireless communication and hardware design. Thereby, limited groups (including companies) took this path, despite of its importance. Table 7.1 lists existing Massive MIMO test-beds that have been reported in public. Their corresponding main features are tabulated as well. LuMaMi test-bed [17] developed at Lund University (Fig. 7.2) is the first 100-antenna test-bed that demonstrates real-time 20 MHz bandwidth transmission.

In summary, many critical open problems have to be investigated and addressed from view points of circuit-and-system design, in order to bring the Massive MIMO technology from theory to practice. This is important for efficient real-life implementation, standardization, and future commercial deployment. These new implementation-related challenges that have been uncovered by Massive MIMO include, but not limited to

- The challenge of dealing with large amount of low-precision analog front-end components, e.g., the solution for time-frequency synchronization and phase coherence of base station RF chains.
- Low-latency and low-power signal processing that supports high mobility (especially for TDD systems) and achieves high total energy efficiency.
- Efficient baseband architecture with optimized processing distribution and interconnection for handling huge amount of data that has to be processed in real-time.

Fig. 7.2 LuMaMi: Lund Massive MIMO Test-bed [17]

The remainder of this chapter discusses challenges in efficient implementation of digital baseband processing, especially in the frame of the reconfigurable architecture for sufficing flexibility requirement in multi-mode operation, multi-algorithm switching, and future system evolution.

7.2 Massive MIMO Baseband Processing

Despite its potential to revolutionize 5G wireless communication networks, Massive MIMO poses critical challenges on hardware realizations, especially on the efficient VLSI implementation of baseband systems (at the base station side). In Massive MIMO, the amount of computations needed has been tremendously scaled up with the number of base station antennas. For instance, the number of elements in a channel matrix is 1000 for a 100-antenna 10-UE setup, i.e., 100× of existing MIMO systems. This change in quantity may lead to quality change in implementation strategies. Combined with 5G requirements of much higher processing throughput and lower latency, much research effort has to be invested on hardware realization issues to achieve efficiency in both silicon area and power consumption. To assistant efficient hardware architecture design, it is crucial to understand the computational operations to be mapped. In this section, we give an overview of the required baseband signal processing in Massive MIMO systems and profile the corresponding processing characteristics.

7.2.1 Baseband Processing Overview

Figure 7.3 shows a simplified block diagram of an OFDM-based TDD Massive MIMO system (at the base station side). With OFDM, system provides good compatibility to existing wireless standards, like 3GPP LTE and LTE-A. TDD is currently the most promising (at least with reasonable implementation costs) technique for realizing duplex Massive MIMO. Nevertheless, the key signal processing techniques discussed here can be extended and applied to other Massive MIMO formats. For each Rx chain in the system, the received RF signals are digitized, followed by analog front-end calibration and time-frequency synchronization, where imperfections in analog components like frequency offset and I/Q imbalance are compensated. From the synchronized data, the Cyclic prefix (CP) is removed, followed by OFDM (using FFT) demodulation and guard-band removal. These OFDM symbols contain the superposition of transmitted signals of all users.

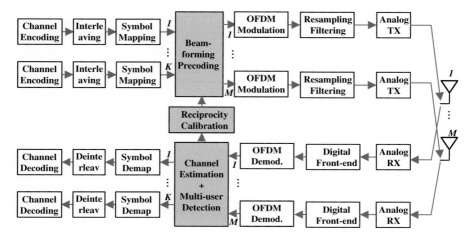

Fig. 7.3 Simplified block diagram of an OFDM-based Massive MIMO system, with M base station antennas and K UE

The frequency-domain signal of each chain is combined and processed by the MIMO detector. Using the channel matrix estimated from uplink pilots, the MIMO detector cancels interference and detects frequency-domain symbols from each UE. The detected symbols are sent to bit-level processing, including de-interleaving and decoding. The downlink baseband processing is basically the reverse processing to uplink, with a unique TDD channel calibration and beam-forming pre-coding. In the following, tree key signal processing blocks in Massive MIMO, i.e., uplink multi-user detection, downlink beam-forming pre-coding, and reciprocity calibration, are introduced.

7.2.2 Uplink Multi-User Detection

Considering an uplink Massive MIMO system with M base station antennas and K single-antenna UE, the received $M \times 1$ complex signal vector is given by

$$r = Hs + n, \tag{7.1}$$

where s is the $K \times 1$ transmitted vector, given that UE are perfectly synchronized. The component in s is taken independently from a set of Gray mapped constellation points. n is the vector of i.i.d. Gaussian noise samples with mean zero and variance N_0 per complex entry, and H denotes the $M \times K$ uplink channel matrix.

The task of uplink multi-user detection is to recover the transmitted vector s given the channel matrix estimation $\hat{H}$ and received vector r. In traditional MIMO systems, many detection algorithms have been discussed, including ZF, MMSE, sphere decoder [5], Markov chain Monte Carlo simulation, etc. As aforementioned, the physically large array in Massive MIMO system experiences different prorogation properties than existing MIMO systems. One of the properties that can be leveraged is that, with massive service antennas, column vectors of the channel matrix H are asymptotically orthogonal, i.e., $H^H H$ becomes diagonally dominated. As a result, linear detection is good enough to provide near-optimal performance, e.g.,

$$\hat{s} = Dr, \tag{7.2}$$

where

$$D = H^H \tag{7.3}$$

for Maximum-ratio combining (MRC) detection (also known as matched filtering) and

$$D = (H^H H + \alpha I)^{-1} H^H \tag{7.4}$$

for regularized ZF detection. Here α is the parameter considering the impact of background noise and unknown user interference.

Worth mentioning is that linear processing in Massive MIMO is not necessarily low-complexity. For instance, the complexity of direct channel matrix inversion for ZF detection is in the order of $\mathcal{O}(10^4)$ for $M = 100$ and $K = 10$.

7.2.3   Downlink Beam-Forming Pre-coding

The equivalent complex-valued baseband model for a Massive MIMO downlink can be expressed as

$$s = Gz + w, \tag{7.5}$$

where s is the $K \times 1$ received vector with each element corresponding to a UE, z is the $M \times 1$ transmitted vector, generated by beam-forming pre-coding, w is the vector of i.i.d. Gaussian noise samples, and G denotes the $K \times M$ downlink channel matrix.

It is generally agreed that wireless propagation channel is reciprocal, e.g.,

$$G^T = H. \tag{7.6}$$

Thereby, uplink channel estimation can be used for downlink beam-forming given that analog impairments can be calibrated out of the system. The task of downlink pre-coding is to "beam-form" the transmitted signal in a way that each user is able to receive almost interference-free signal. As a consequence, signal processing at the UE side can be significantly simplified. Many schemes exist for pre-coding the downlink signal x, including linear and non-linear techniques. Similar to uplink detection, linear processing is generally good enough given the huge degree-of-freedom in service antennas. Linear pre-coding generates z as

$$z = Cx, \tag{7.7}$$

where x is the $K \times 1$ signal vector, corresponding to K users. The pre-coding matrix C has the form of

$$C = G^H \tag{7.8}$$

and

$$C = G^H(GG^H + \alpha I)^{-1} \tag{7.9}$$

for Maximum Ratio Transmission (MRT) and regularized ZF pre-coding, respectively.

Other techniques like Constant Envelope (CE) pre-coding are being investigated to relax the requirement of RF front-end in Massive MIMO systems. The motivation behind the strict constraint on the amplitude of radiated signals (change only in phase) is the use of power-efficient linear power amplifiers, which can reduce hardware cost significantly in comparison to those used in existing cellular systems.

It should be pointed out that responses of analog components need to be taken into account in practical system design, since they may destroy the nice reciprocal property due to circuit variations, manufacturing process, voltage supply, environment temperature, etc. Thereby, differences in analog chains between uplink and downlink have to be estimated and compensated, commonly referred to as TDD channel reciprocity calibration [15].

7.3 New Challenges in Reconfigurable Architecture Design

In this section, we look into the reconfigurable baseband processor for Massive MIMO systems, based on the introduction in Sect. 7.2. Designing an efficient reconfigurable architecture for Massive MIMO is a new and challenging task. The orders-of-magnitude increase in the number of data paths requires revolutionary re-thinking beyond existing architectures. The purpose of this section is not to describe or propose an architecture. Instead, we would like to open up discussion on this interesting topic, by first analyzing unique features in Massive MIMO baseband processing and then looking into architecture design challenges corresponding to these new features. We believe the discussion in this session will serve as a design guideline for efficient reconfigurable architecture design in Massive MIMO applications.

7.3.1 Computational Complexity

To facilitate the design of baseband processor, it is important to analyze and profile computational complexity. Table 7.2 lists the number of operations involved in each processing block in a typical OFDM-based Massive MIMO baseband system (Fig. 7.3). In baseband processors, multipliers are commonly the dominating arithmetic function units, in terms of both silicon area and power consumption. Thereby, the presented complexity analysis only includes the number of complex-valued multiplications. Other complicated but rare operations, like division and square-root, can be accelerated and attached as co-processors (similar to the case study presented in Chap. 6).

The number of operations in Table 7.2 is parameterized based on the number of base station antennas M, the number of single-antenna UE K, and system bandwidth (affecting the number of data subcarriers N_c and FFT/IFFT size N_{FFT}). As can be seen, the number of operations is in the order of $\mathcal{O}(10^7)$ for a 100-antenna 10-UE Massive MIMO system with 20 MHz bandwidth. Moreover, this large number of operations have to be performed with high throughput and low latency, especially in TDD systems. One potential technique to meet this critical design requirement

Table 7.2 Computational complexity for Massive MIMO baseband processing

Processing block	Algorithm	Multiplication
FFT		$M \times N_{\text{FFT}} \times \log_2(N_{\text{FFT}})$
Channel estimation		$M \times N_c$
Channel interpolation	Linear interpolation	$2M \times N_c$
Data detection pre-processing	ZF	$2 \times N_c \times K^2 \times M + 2 \times K^3$
	MRC	$N_c \times K \times M$
Data detection		$N_c \times K \times M$
Reciprocity calibration		$N_c \times K \times M$
Data pre-coding pre-processing	ZF	$2 \times N_c \times K^2 \times M + 2 \times K^3$
	MRC	$3 \times N_c \times K \times M$
Data pre-coding		$N_c \times K \times M$
IFFT		$M \times N_{\text{FFT}} \times \log_2(N_{\text{FFT}})$
Total for minimum requirement	MRC (uplink)+MRC (downlink)	$M \times (5N_{\text{FFT}} \times \log_2(N_{\text{FFT}}) + 3N_c + 8KN_c) \approx 2.1 \times 10^7$
Total for maximum requirement	ZF (uplink)+ZF (downlink)	$M \times (5N_{\text{FFT}} \times \log_2(N_{\text{FFT}}) + 3N_c + 5KN_c + 4K^2N_c) \approx 6.7 \times 10^7$

is massive parallel computing, e.g., using SIMD architecture. In the context of Massive MIMO, Data-level parallelism (DLP) can be exploited in both spatial and frequency domain, e.g., operations for different receiver chains and data subcarriers.

7.3.2 Processing Distribution

In addition to DLP, processing distribution is another important design aspect in Massive MIMO systems. With orders-of-magnitude increase in data paths, processing distribution can be used to reduce overhead in data shuffling network and memory sub-system. Moreover, with newly emerged wireless communication technologies, e.g., distributed antenna system, this aspect will become even more important.

Basically, there are two types of processing distributions, centralized and distributed processing. In centralized systems, data from/to all RF chains are aggregated into one main processor, which handles all the required computations. The advantage of centralized processing is the capability of performing advanced algorithms for achieving better system performance. The price to pay is the overhead of data transmission and shuffling. In contrast, distributed processing localizes operations close to data producer/consumer, aiming to reduce data movements. Comparing the two distribution strategies, it is difficult to say which one is better. The optimal decision is highly dependent on algorithm selections and design specification.

Fig. 7.4 Processing
distribution for Massive
MIMO pre-coding

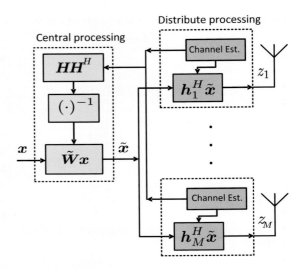

Figure 7.4 shows an example of processing distribution strategy for Massive
MIMO downlink beam-forming pre-coding. For MRC pre-coding,

$$z = H^H x, \tag{7.10}$$

channel estimation and pre-coding may be localized close to antennas, i.e., using
distributed processing, in order to reduce the amount of data shuffling and on-chip
network bandwidth. On the other side, hybrid strategy may be adopted for ZF pre-
coding,

$$z = H^H (HH^H)^{-1} x. \tag{7.11}$$

For instance, ZF process can be divided into two steps:

$$\tilde{x} = (HH^H)^{-1} x \tag{7.12}$$

and

$$z = H^H \tilde{x}, \tag{7.13}$$

where $\tilde{x}$ is computed centrally and $z = H^H \tilde{x}$ is performed in a distributed manner.

7.3.3 Spatial-Domain Selectivity

Radio systems experience highly dynamic operating conditions, where both propagation environments and QoS requirements vary within huge ranges. Conventional circuits for radio communication applications are conservative and are often designed to guarantee reliable worst-case processing, thus consuming more power than necessary in most situations. On the other side, adaptive computing is a promising approach that significantly reduces power consumption by performing power-efficient and good-enough processing, based on the current channel condition and application-specified QoS requirements. This smartness in processing requires flexible hardware support.

Generally, wireless channel is selective in three dimensions, namely time, frequency, and space. Many studies have been conducted to exploit the selectivity in time and frequency domain, while the diversity in spatial domain has not been well studied until the invention of MIMO technology. Equipped with a large number of service antennas, Massive MIMO systems potentially provide much more spatial selectivity in the propagation channel. The scenario can be either different number of users being served or variable geographical distributions of connected users. For example, a large number of users are close to each other in a stadium, while users are separated in rural areas. In the latter case, the channel of each user is close-to-orthogonal. Therefore, linear processing (e.g., matched filter, ZF, and MMSE) is good enough to provide near-optimal performance. While in the former scenario, the equivalent channels are highly correlated in spatial domain, requiring more advanced non-linear processing to maintain system performance. Other examples include the different number of served users between day-time and night. In this case, adaptive and reconfigurable computing can be utilized to enable power reduction. For instance, the antenna selection technique can be applied to reduce the number of active antennas when less users are connected, while preserving system performance at a certain required level. By switching off antennas, signal processing can be simplified because of the reduced number of data paths.

We believe much more adaptive signal processing algorithms will be introduced to fully harvesting the available spatial-domain selectivity in Massive MIMO, which in turn poses tremendous flexibility requirements and design challenges to baseband processors.

7.4 Summary

This chapter introduces an emerging wireless communication technology, Massive MIMO, for the coming 5G network. Besides the basic concept of the Massive MIMO, its main advantages over existing systems are discussed, including the prorogation feature with large antenna array, the capability of improving both spectral and transmitted energy efficiency, and the benefit of extensive use of low-cost electric components. To bring this promising technique from theory to practice, state-of-the-art research progress in the area of Massive MIMO is recapped. Additionally, unique operations required for digital signal processing in Massive MIMO baseband (at the base station side) are profiled and analyzed. In Massive MIMO, beam-forming pre-coding and multi-user detection are conducted with large-scale channel matrices. Channel reciprocity calibration is crucial to enable practice TDD Massive MIMO transmission. Based on the operation analysis, we identify tree key features in Massive MIMO baseband processing that have critical impacts on reconfigurable hardware design. Besides orders of magnitudes increase in computational complexity, system architects may consider optimal processing distribution to relax the requirement of on-chip network and memory sub-system. Moreover, more flexibility is needed to explore the spatial-domain selectivity in Massive MIMO systems.

References

1. X. Gao, F. Tufvesson, O. Edfors, F. Rusek, Measured propagation characteristics for very-large MIMO at 2.6 GHz, in *Asilomar Conference on Signals, Systems and Computers* (2012), pp. 295–299
2. S. Hong et al., Applications of self-interference cancellation in 5G and beyond. IEEE Commun. Mag. **52**(2), 114–121 (2014)
3. Huawei, 5G, a technology vision, 2013. http://www.huawei.com/5gwhitepaper
4. E.G. Larsson, O. Edfors, F. Tufvesson, T.L Marzetta, Massive MIMO for next generation wireless systems. IEEE Commun. Mag. **52**(2), 186–195 (2014)
5. L. Liu, F. Ye, X. Ma, T. Zhang, J. Ren, A 1.1-Gb/s 115-pJ/bit configurable MIMO detector using 0.13-μCMOS technology. IEEE Trans. Circuits Syst. Express Briefs **57**(9), 701–705 (2010)
6. N. Michailow, et al., Generalized frequency division multiplexing: a flexible multi-carrier modulation scheme for 5th generation cellular networks, in *German Microwave Conference (GeMiC)*, 2012
7. P. Mogensen, et al., 5G small cell optimized radio design, in *IEEE Globecom Workshops* (2013), pp. 111–116
8. Y.H. Nam, et al., Full-dimension MIMO (FD-MIMO) for next generation cellular technology. IEEE Commun. Mag. **51**(6), 172–179 (2013)
9. H.Q. Ngo, E.G. Larsson, T.L. Marzetta, Energy and spectral efficiency of very large multiuser MIMO systems. IEEE Trans. Commun. **61**(4), 1436–1449 (2013)
10. H.Q. Ngo, et al., Uplink performance analysis of multicell MU-SIMO systems with ZF receivers. IEEE Trans. Veh. Technol. **62**(9), 4471–4483 (2013)
11. H. Prabhu, J. Rodrigues, O. Edfors, F. Rusek, Approximate matrix inverse computations for very-large MIMO and applications to linear pre-coding systems, in *IEEE Wireless Communications and Networking Conference (WCNC)* (2013), pp. 2710–2715
12. H. Prabhu, et al., A low-complex peak-to-average power reduction scheme for OFDM based Massive MIMO systems, in *IEEE International Symposium on Communications, Control and Signal Processing (ISCCSP)* (2014), pp. 114–117
13. H. Prabhu, F. Rusek, J. Rodrigues, O. Edfors, High throughput constant envelope pre-coder for Massive MIMO systems, in *IEEE International Symposium on Circuits and Systems (ISCAS)* (2015), pp. 1502–1505

14. T. Rappaport, et al., Millimeter wave mobile communications for 5G cellular: it will work! IEEE Access **1**, 335–349 (2013)
15. R. Rogalin, et al., Scalable synchronization and reciprocity calibration for distributed multiuser MIMO. IEEE Trans. Wireless Commun. **13**(4), 1815–1831 (2015)
16. F. Rusek, et al., Scaling up MIMO: opportunities and challenges with very large arrays. IEEE Signal Process. Mag. **30**(1), 40–60 (2013)
17. J. Vieira, et al., A flexible 100-antenna testbed for Massive MIMO, in *IEEE Globecom Workshop-Massive MIMO: From Theory to Practice*, 2014, pp. 287–293
18. B. Yin, et al., Implementation trade-offs for linear detection in large-scale MIMO systems, in *IEEE International Conference on Acoustics, Speech and Signal Processing (ICASSP)* (2013), pp. 2679–2683
19. B. Yin, et al., A 3.8 Gb/s large-scale MIMO detector for 3GPP LTE-advanced, in *IEEE International Conference on Acoustics, Speech and Signal Processing (ICASSP)* (2014), pp. 3879–3883

Chapter 8
Conclusion

Coarse-grained reconfigurable architectures (CGRAs) emerge as a new class of hardware platforms, designed to bridge the gap of computational performance, hardware efficiency, and flexibility among conventional architectures such as Application-specific integrated circuits (ASICs), Field-programmable gate arrays (FPGAs), and Digital signal processors (DSPs). The strength of CGRAs lies in the capability of allocating hardware resources dynamically to accomplish current computational demands. In addition, hardware efficiency with respect to area and power consumption is substantially improved in comparison to FPGAs, thanks to the word-level data manipulations.

In this book, a dynamically reconfigurable cell array architecture is introduced, developed, and verified in silicon with a primary focus on digital baseband processing in wireless communication. The presented cell array architecture is constructed from an array of processing and memory cells interconnected through a hierarchical easily-scalable on-chip network. High hardware efficiency is attained by conducting algorithm-architecture, hardware-software, and processing–memory co-design. The performance and flexibility of the cell array are demonstrated through two case studies.

In the first study, the cell array is designed to process multiple radio standards concurrently, aiming to demonstrate the flexibility of the architecture and evaluate the control overhead, in terms of clock cycles and area consumption, of hardware reconfigurations. With a 2×2 cell array, three contemporary wireless communication standards are supported and two independent data streams from any of the three standards can be processed concurrently. Depending on the number of receiving data streams, the cell array dynamically adjusts its underlying hardware resources to maximize hardware usage for achieving either high computational accuracy or processing concurrence. In addition to resource sharing among multiple radio standards, hardware flexibility is demonstrated by mapping different algorithms onto the same platform after chip fabrication. The adoption of a newly developed algorithm

© Springer International Publishing Switzerland 2016
C. Zhang et al., *Heterogeneous Reconfigurable Processors for Real-Time Baseband Processing*, DOI 10.1007/978-3-319-24004-6_8

extends the coverage of standards to be supported. Thanks to the employed in-cell configuration scheme, run-time context switching between different operation scenarios requires at most 11 clock cycles, which correspond to the configuration time of $\sim$34 nS at 320 MHz. Implementation results show that the adoption of the cell array in a digital front-end receiver requires only about 16 % area overhead in comparison to its ASIC counterpart.

The second study deals with multi-task processing, aiming to demonstrate the flexibility and real-time processing capability of the cell array as well as to evaluate the area and energy efficiency. To this end, three crucial and compute-intensive baseband processing blocks in 4×4 MIMO-OFDM systems, namely channel estimation, pre-processing, and symbol detection, are mapped onto the cell array. Benefiting from algorithm-architecture co-design, the cell array is capable of processing all the target tasks in real-time for a 20 MHz 64-QAM 3GPP LTE-A downlink. On the algorithm side, most of the operations are unified to vector-level, which enables extensive parallel processing and resource sharing for attaining high hardware efficiency. On the architecture side, the cell array is extended with extensive vector computing capabilities, including vector-enhanced SIMD cores and VLIW-style multi-stage computation chain. This is done in order to achieve low-latency high-throughput vector computing and reduce register/memory access for loading and storing intermediate results. In addition, flexible memory access schemes are adopted to relieve processing cores from non-computational address manipulations. Implementation results show that the presented cell array outperforms related programmable platforms by up to 6 orders of magnitude in energy efficiency, and achieves similar level of efficiency to that of ASICs in terms of area and energy.

This book also initiates the study and discussion on reconfigurable architecture design for Massive MIMO baseband processing applications. The orders-of-magnitude increase in data path in Massive MIMO systems requires extension and modification to the cell array and (SIMD) architecture to support its baseband processing. How to further explore the concept of Data-level parallelism (DLP) in both frequency- and spatial-domain is one of the key design topics.

In conclusion, the CGRA-based cell array demonstrates a good design trade-off between the contradictory requirements of flexibility, performance, and hardware efficiency. Thus, it is a promising and feasible solution to bridge the huge gap between conventional platforms. Looking forward, adopting the cell array in a wide range of applications is a natural continuation. However, this requires a series of system-level exploration tools to model, simulate and evaluate the use of the platform as well as application mapping tools to automate task profiling, scheduling, mapping, and compilation process. Despite the system-level developments to be explored, the future of the reconfigurable architecture is certainly bright.

Appendix A
Dataflow Processor Architecture

This appendix includes some detailed hardware development of the 2×2 cell array presented in Chap. 5. The cell array consists of two dataflow processors and two memory cells. The dataflow processors are Reduced instruction set computing (RISC) cores with extended computational units in both "instruction decode" and "write back" stage. Each processor contains 19 general-purpose 16-bit registers and uses a 48-bit fixed-length instruction set. Some of the key features of the processors are:

- Single instruction multiple data (SIMD)-like operation
- Run-time control and data path configuration
- Conditional instruction execution
- Single-cycle delayed branch
- Zero-delay inner loop control
- Direct I/O port addressing and multi-port data streaming
- In-cell Resource cell (RC) supervision and configuration

Figures A.1, A.2, and A.3 and Tables A.1 and A.2 present the instruction set of the dataflow processor. Figure A.4 and Tables A.3, A.4, A.5, and A.6 illustrate the data arrangement blocks and list the configuration set of each pipeline stage. Figures A.5, A.6, and A.7 present the configuration generation tool developed in-house for the 2×2 cell array. Tables A.7 and A.8 describe user commands in the UART and the MATLAB interface, respectively, for controlling the cell array at run-time.

© Springer International Publishing Switzerland 2016
C. Zhang et al., *Heterogeneous Reconfigurable Processors for Real-Time Baseband Processing*, DOI 10.1007/978-3-319-24004-6

Mnemonics	Operands	Type	Description	Operation
NOP		A / D	No operation/Configuration updates	–
GID	Imm	E	Global IO TX port destination ID write	$GO.dst <- Imm
END	Imm	E	Stop execution and return ending code	SEND_CODE <- Imm
ILC	S1, D1	D	Inner loop control with register	$ILP <- $PC + 1; D1 <= $ILC; $ILC <- S1
ILCI	Imm	E	Inner loop control with imm. Infinite loop control when imm. is all "1"	$ILP <- $PC + 1; $ILC <- Imm
LDR	D0, D1, S0, S1	A	MC (RAM) data read request	D0 <- S0 & S1
LDRI	D0, Imm	C		D0 <- Imm
STR	D0, D1, S0, S1	A	MC (RAM) data write request	D0 <- S0 & S1
STRI	D0, Imm	C		D0 <- Imm
CONFI	D0, Imm	C	Resource cell configuration	D0 <- Imm
BR		D	Branch register (procedure return)	$PC <- $R_link
B	S0, S1, Imm	B	Branch relative on condition	$PC <- $PC + SXT(Imm)
BL	S0, S1, Imm	B	Branch and Link (procedure call) on condition	$PC <- $PC + SXT(Imm); $R_link <- $PC + 1

Field (bits 47 … 0)

Mnemonic	47–44	43–38	37 (I)	37 (P)	36–34	33–31	30–18	17–10	9–6	5–3	2–0
(hdr)	Cond.	OpCode	I	P	Index	Add.	D0	S0	S1		D1
(hdr)	Cond.	OpCode	I	P	Index	Add.	Imm.		S0	S1	Imm.
(hdr)	Cond.	OpCode	I	P		D0	32-bit memory data read/write request				
(hdr)	Cond.	OpCode	I	P	Index	Add.	21-bit configuration				D1
(hdr)	Cond.	OpCode	I	P	Index	Add.	21-bit configuration				D1
NOP	Cond.	0 0 0 0 0 0	0	P	Index	Add.	21-bit configuration				
GID	Cond.	0 0 0 0 1 1	0	P	Index	Add.	21-bit configuration				
END	Cond.	0 0 1 0 1 0	0	P	Index	Add.	21-bit configuration				
ILC	Cond.	0 0 0 1 1 0	1	P	Index	Add.	21-bit configuration	S1			D1
ILCI	Cond.	0 0 1 1 1 1	1	P	Index	Add.	21-bit configuration	Imm.			D1
LDR	Cond.	0 0 1 0 0 1	0	0	D0		32-bit read request	S0	S1		D1
LDRI	Cond.	0 1 0 0 1 0	0	0	D0		32-bit read request	D0			D1
STR	Cond.	0 1 0 1 0 1	0	0	D0		32-bit write request	S0	S1		
STRI	Cond.	0 1 1 1 0 1	0	0	D0		32-bit write request	D0			
CONFI	Cond.	0 1 0 1 1 0	0	P	Index	Add.	21-bit configuration				
BR	Cond.	0 0 0 0 0 0	0	U		Imm.		$R_link			
B	Cond.	0 1 0 1 0 0	0	0		Imm.		S0	S1		
BL	Cond.	0 1 0 1 0 0	0	U		Imm.	$R_link	S0	S1		

Fig. A.1 Instruction set of the dataflow processor, control-related operations

Mnemonics	Operands	Type	Description	Operation	Cond. (47–42)	OpCode1 (41–38)	OpCode2 (37–36)	35	U (34)	L (33)	Field (32–25)	ID (24)	EXE (23–21)	WD (20–18)	D0 (17–15)	S0 (12–10)	S1 / Imm (9–5)	D1 (2–0)
ADD	D0, D1, S0, S1	F	Data add between registers	D0 ← S0 + S1	Cond.	1 0 0 0		0	U	L	V0 V1 A0 A1 R0 R1	ID	EXE	WD	D0	S0	S1	D1
ADDI	D0, S0, Imm	G	Add register data and imm.	D0 ← S0 + SXT(Imm)	Cond.	1 0 0 0		1	U	L	V0 V1 A0 A1 R0 R1	ID	EXE	WD	D0	S0	Imm.	D1
ADC	D0, D1, S0, S1	F	Data add between registers	D0 ← S0 + S1 + Carry	Cond.	1 0 0 1		0	U	L	V0 V1 A0 A1 R0 R1	ID	EXE	WD	D0	S0	S1	D1
ADCI	D0, S0, Imm	G	Add register data and imm.	D0 ← S0 + SXT(Imm) + Carry	Cond.	1 0 0 1		1	U	L	V0 V1 A0 A1 R0 R1	ID	EXE	WD	D0	S0	Imm.	D1
SUB	D0, D1, S0, S1	F	Data subtract between registers	D0 ← S0 - S1	Cond.	1 0 1 0		0	U	L	V0 V1 A0 A1 R0 R1	ID	EXE	WD	D0	S0	S1	D1
SBC	D0, D1, S0, S1	F	Data subtract between registers	D0 ← S0 - S1 - Carry	Cond.	1 0 1 1		0	U	L	V0 V1 A0 A1 R0 R1	ID	EXE	WD	D0	S0	S1	D1
ADS	D0, D1, S0, S1	F	Data add and subtract between registers	D0.lo ← S0.lo + S1.lo; D0.hi ← S0.hi - S1.hi	Cond.	1 0 1 0 0		0	U	L	V0 V1 1 1 R0 R1	ID	EXE	WD	D0	S0	S1	D1
ADSC	D0, D1, S0, S1	F	Data add and subtract between registers	D0.lo ← S0.lo + S1.lo + Carry; D0.hi ← S0.hi - S1.hi - Carry	Cond.	1 0 1 0 1		0	U	L	V0 V1 1 1 R0 R1	ID	EXE	WD	D0	S0	S1	
MUL	D0, D1, S0, S1	F	Multiply register data and imm.	D0 ← S0 * S1	Cond.	1 0 1 1	0 0 0 0	0	U	L	V0 V1 A0 A1 R0 R1	ID	EXE	WD	D0	S0	S1	D1
MULI	D0, S0, Imm	G	Data multiplication between registers	D0 ← S0 * SXT(Imm)	Cond.	1 0 1 1	0 0 0 1	0	U	L	V0 V1 A0 A1 R0 R1	ID	EXE	WD	D0	S0	Imm.	D1
CMP	D0, D1, S0, S1	F	Compare date between registers	As SUB without writing result(s)	Cond.	1 1 0 0 0		0	U	L	V0 V1 A0 A1 R0 R1	ID	EXE	WD	D0	S0	S1	D1
CMPI	D0, S0, Imm	G	Compare register data and imm.	As ADDI without writing result(s)	Cond.	1 1 0 0		1	U	L	V0 V1 A0 A1 R0 R1	ID	EXE	WD	D0	S0	Imm.	D1
TST	D0, D1, S0, S1	F	Register data test	As AND without writing result(s)	Cond.	1 1 0 1 0		0	U	L	V0 V1 A0 A1 R0 R1	ID	EXE	WD	D0	S0	S1	D1
TSTI	D0, S0, Imm	G	Imm. data test	As AND without writing result	Cond.	1 1 0 1		1	U	L	Imm. - High	ID	EXE	WD	D0	S0	Imm. - Low	D1
TEQ	D0, D1, S0, S1	F	Register data equality test	As XOR without writing result(s)	Cond.	1 1 1 0 1		0	U	L	Imm. - High	ID	EXE	WD	D0	S0	Imm. - Low	D1
TEQI	D0, S0, Imm	G	Imm. data equality test	As XOR without writing result	Cond.	1 1 1 0		1	U	L	Imm. - High	ID	EXE	WD	D0	S0	Imm. - Low	D1
AND	D0, D1, S0, S1	F	Logical AND two register data	D0 ← S0 & S1	Cond.	1 1 1 1 0		0	U	L	Imm. - High	ID	EXE	WD	D0	S0	Imm. - Low	D1
ANDI	D0, S0, Imm	G	Logical AND register data with imm.	D0 ← S0 & Imm	Cond.	1 1 1 1		1	U	L	V0 V1 A0 A1 R0 R1	ID	EXE	WD	D0	S0	Imm. - Low	D1
OR	D0, D1, S0, S1	F	Logical OR two register data	D0 ← S0	S1	Cond.	1 1 1 1	OpCode2	0	U	L	V0 V1 A0 A1 R0 R1	ID	EXE	WD	D0	S0	S1	D1
ORI	D0, S0, Imm	G	Logical OR register data with imm.	D0 ← S0	Imm	Cond.	1 1 1 1		1	U	L	Imm. - High	ID	EXE	WD	D0	S0	Imm. - Low	D1
XOR	D0, D1, S0, S1	F	Logical XOR two register data	D0 ← S0 ⊕ S1	Cond.	1 1 1 1		0	U	L	V0 V1 A0 A1 R0 R1	ID	EXE	WD	D0	S0	S1	D1
XORI	D0, S0, Imm	G	Logical XOR register data with imm.	D0 ← S0 ⊕ Imm	Cond.	1 1 1 1		1	U	L	Imm. - High	ID	EXE	WD	D0	S0	Imm. - Low	D1
MOV	D0, D1, S0, S1	F	Data move between registers	D0 ← S0	Cond.	1 1 1 1		0	U	L	V0 V1 A0 A1 R0 R1	ID	EXE	WD	D0	S0	S1	D1
MOVI	D0, S0D1, Imm	G	Move imm. to a register	D0, D1 ← imm.	Cond.	1 1 1 1 1		1	U	L	Imm. - High	ID	EXE	WD	D1	S0	Imm. - Low	D1

Fig. A.2 Instruction set of the dataflow processor, arithmetic and logic operations

Operation	Description	Direction	Data package
			31 30 29 28 27 26 25 24 / 23 22 21 20 19 18 17 16 15 14 13 12 11 10 / 9-3 / 2 / 1 / 0
Program download Header	Package header for downloading program into ProGram Memory (PGM). Address 0 is reserved for the Control register, therefore instructions start from address 1.	Host -> PC	Count / Starting address / Conditional configuration address / CC / P / Write
Program download	Instruction to be downloaded into ProGram Memory. This is a consecutive operation of the "Program download header".	Host -> PC	Instruction
PC counter update	Update PC counter starting address, to select program section inside ProGram Memory (PGM).	Host -> PC	0 / Starting address / Conditional configuration address / CC / P / Write
Control register update	Update Control register to control the operations of the Processor Cell.	Host -> PC	Stop / Run to / Step / Reset / Pause / Start / Ending address / Conditional configuration address / CC / C / Write
Control register read request	Control register status reading request.	Host -> PC	CC / C / Read
Control register read data 1	Sending back Control register status, data package 1. This is a consecutive operation of "Control register read request".	PC -> Host	Stop / Run to / Step / Reset / Pause / Start / Ending address / Current PC counter
Control register read data 2	Sending back Control register status, data package 2. This is a consecutive operation of "Control register read data 1".	PC -> Host	Instruction END code

Fig. A.3 Control instruction set of the dataflow processor

Table A.1 Conditional field
of the instruction set

Opcode	ALU-based condition	Co-ALU-based condition
0000	Equal	Equal
0001	Not equal	Not equal
0010	Negative	Less than
0011	Non-negative	Less than or equal
0100	Carry out	Greater than
0101	Non-carry out	Greater than or equal
0110	Overflow	Positive operand "a"
0111	Non-overflow	Negative operand "a"
1000	Less than	Positive operand "b"
1001	Less than or equal	Negative operand "b"
1010	Greater than	Positive operand "c"
1011	Greater than or equal	Negative operand "c"
1100	Reserved	Positive operand "d"
1101	Reserved	Negative operand "d"
1110	Reserved	Branch not taken
1111	Always	Always

Table A.2 Summary of
register address

Register	Address	Register	Address
General-purpose registers		*Special purpose registers*	
$1	00100	ZERO	00000
$2	00101	PC	00001
$3	00110	Link	00010
$4	00111	Stack	00011
$5	01000		
$6	01001	*Hierarchical IO registers*	
$7	01010	G0	10111
$8	01011		
$9	01100	*Local IO registers*	
$10	01101	L0	11000
$11	01110	L1	11001
$12	01111	L2	11010
$13	10000	L3	11011
$14	10001	L4	11100
$15	10010	L5	11101
$16	10011	L6	11110
$17	10100	L7	11111
$18	10101		
$19	10110		

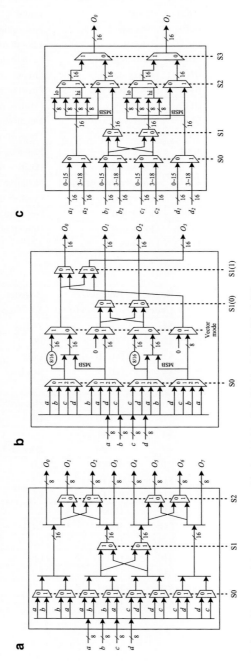

Fig. A.4 Architecture of data arrangement blocks, (**a**) block-I, (**b**) block-II, (**c**) block-III

Table A.3 Configuration set for function units in instruction decoding stage

Bit	Field	Description
20-19	opcode_a	Operation code (Table A.4) for Barrel shifting, operand A
18-17	opcode_b	Operation code (Table A.4) for Barrel shifting, operand B
16	bs_inc_en_a	Enable of automatic shifting increment, operand A
15	bs_en_a	Barrel shifting enable for operand A
14-12	bs_imm_a	Shifting bit count for operand A
11	bs_inc_en_b	Enable of automatic shifting increment, operand B
10	bs_en_b	Barrel shifting enable for operand B
9-7	bs_imm_b	Shifting bit count for operand B
6-3	neg_en	Enable of input data negation (d, c, b, a)
2	sel_alu	ALU status register selection, 0: Co-ALU; 1: ALU
1-0	sel_lane	Processing lane selection

Table A.4 Operation code for barrel shifter

Opcode	Operation
00	Logical shift left (LSL)
01	Logical shift right (LSR)
10	Arithmetic shift right (ASR)
11	ROtate right (ROR)

Table A.5 Configuration set for function units in execution stage

Bit	Field	Description
20-13	mux_1_s0	Control bits for data arrangement block-I, stage 0
12	mux_1_s1	Control bits for data arrangement block-I, stage 1
11-10	mux_1_s2	Control bits for data arrangement block-I, stage 2
9-2	mux_2_s0	Control bits for data arrangement block-II, stage 0
1-0	mux_2_s1	Control bits for data arrangement block-II, stage 1

Table A.6 Configuration set for function units in write back stage

Bit	Field	Description
14-11	add_sub	Accumulator ADD/SUB select (d, c, b, a), 0: addition; 1: subtraction
10-7	mux_3_s0	Control bits for data arrangement block-III, stage 0
6	mux_3_s1	Control bits for data arrangement block-III, stage 1
5-2	mux_3_s2	Control bits for data arrangement block-III, stage 2
1-0	mux_3_s3	Control bits for data arrangement block-III, stage 3

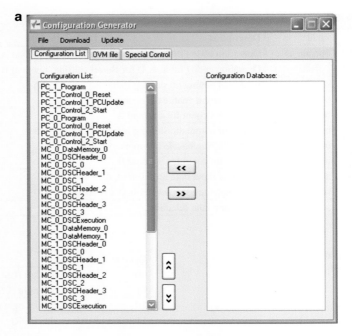

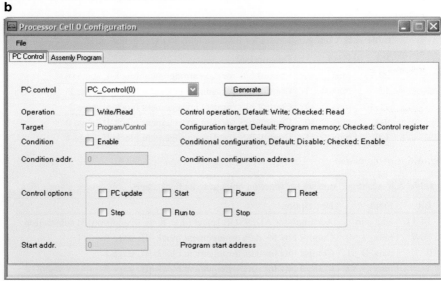

Fig. A.5 Configuration generation tool, (**a**) bit stream generation, (**b**) configuration of a processing cell

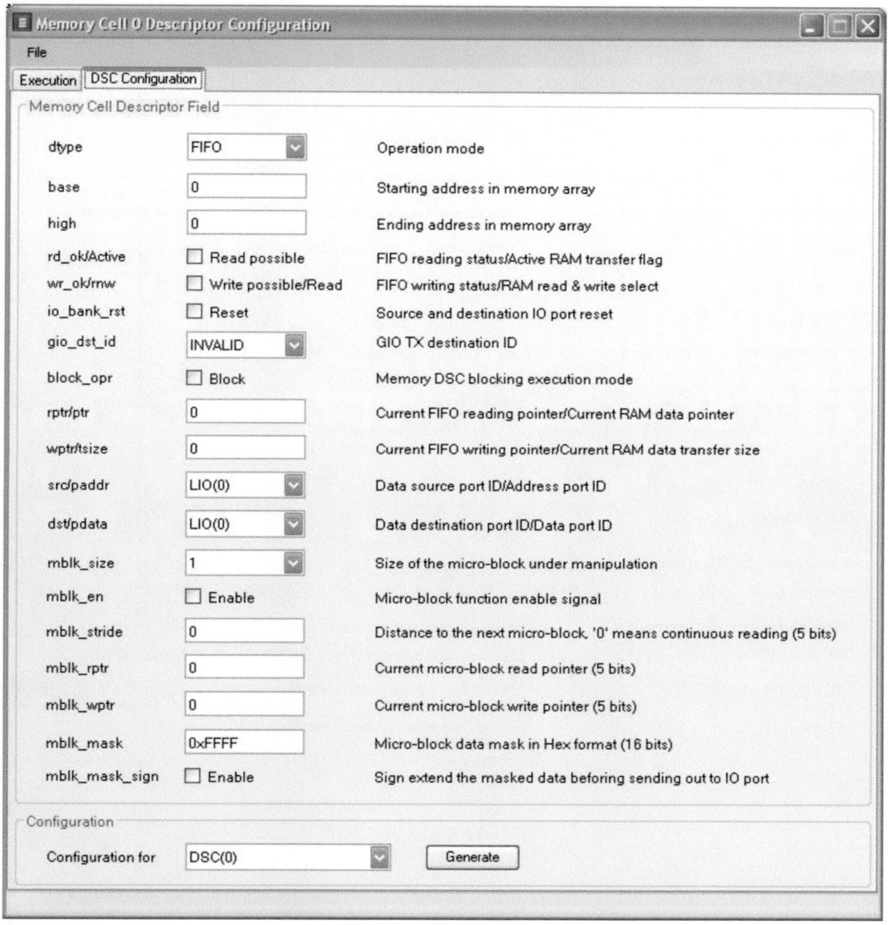

Fig. A.6 Configuration generation tool, descriptor configuration of a memory cell

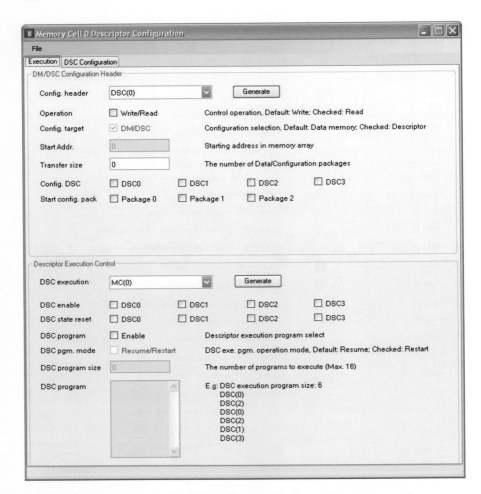

Fig. A.7 Configuration generation tool, configuration header, and descriptor execution program of a memory cell

Table A.7 User commands in UART interface

CMD	Description	Parameter
g	Destination cell ID for hierarchical I/O communication	Resource cell destination ID (number input): 0–4
d	Send data inputs via UART to the selected cell	(a) Data count (number input) (b) Data inputs (string input)
D	Send data inputs via Ethernet to the selected cell	None
i	Send inst./config. packages via UART to the selected cell	(a) Inst. count (number input) (b) Inst./config. inputs (string input)
I	Send inst./config. packages via Ethernet to the selected cell	None
s	Send "start" command to the selected processing cell	None
p	Send "pause" command to the selected processing cell	None
r	Send "reset" command to the selected processing cell	None
e	Send "step" command to the selected processing cell	None
u	Send "run to" command to the selected processing cell	Destination instruction to run to (number input)
o	Send "stop" command to the selected processing cell	None
t	Status tracing of the selected cell	None
c	Send user command to the selected cell	User command (string input)
f	Memory data initialization (zero filling)	Memory cell destination ID (number input): 1, 2
q	Processing cell initialization	Processing cell destination ID (number input): 0, 3
z	Test data set input	None
0	Demo full config. script	None
1	Demo partial config. script	None
h	Command help printout	None

Table A.8 User commands in MATLAB interface

CMD	Description	Parameter
cmd	User command input in UART interface	UART commands
config	Send inst./config. packages from a script file	None
data	Send data inputs from a script file	None
demo	Run a script demo, IEEE 802.11n Sync.	None
rxbuf	Flush UART Rx buffer of the host	None
help	Command help printout	None
exit	Exit user interface in MATLAB	None

Appendix B
Vector Dataflow Processor Architecture

This appendix includes detailed micro-code and instruction set of the vector dataflow processor (Tile-0) in the reconfigurable cell array presented in Chap. 6. The processor is composed of three processing cells for data computations, one memory cell for local data buffering, and a sequencer for control-flow managements. Figure B.1 illustrates seven-stage pipeline of the processor. Note that operations mapped onto two function units in the pre-processing stage can be executed concurrently. Tables B.1, B.2, B.3, B.4, B.5, B.6, B.7, B.8, B.9, B.10, B.11, B.12, and B.13 present the micro-code set for each processing and memory cell and Table B.13 describes the instruction set of the sequencer.

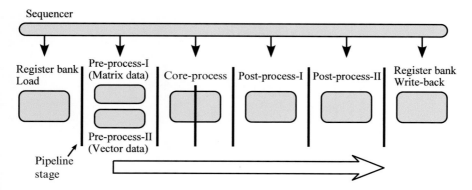

Fig. B.1 Microarchitecture of the vector dataflow processor, a view of pipeline stages

© Springer International Publishing Switzerland 2016
C. Zhang et al., *Heterogeneous Reconfigurable Processors for Real-Time
Baseband Processing*, DOI 10.1007/978-3-319-24004-6

Table B.1 Micro-code set for the data loading stage of register bank

Bit	Division	Field	Description
31-22	Reserved		
21	Matrix data operand	reg_mopa_en	I/O port enable for operand A reading, 0: disable; 1: enable
20		reg_mopa_src	Data source for operand A reading, 0: GPR; 1: I/O
19-18		reg_mopa_idx[a]	I/O port number/GPR index for operand A reading
17		reg_mopb_en	I/O port enable for operand B reading
16		reg_mopb_src	Data source for operand B reading, 0: GPR; 1: I/O
15-14		reg_mopb_idx[a]	I/O port number/GPR index for operand B reading
13	Vector data operand	reg_vop_en	I/O port enable for vector operand reading
12		reg_vop_src	Data source for vector operand reading, 0: GPR; 1: I/O
11-8		reg_vop_idx	I/O port number/GPR index for vector operand reading
7-4	VPR	vpr_idx	Index of Vector permutation register (VPR)
3-0	MMR	mmr_idx	Index of Matrix mask register (MMR)

[a]"Multiple-of-four" addressing (2 LSBs are discarded) for loading matrix data from GPR, e.g., accessing $0, $4, $8, etc.

Table B.2 Micro-code set for pre-processing cell I (matrix data)

Bit	Division	Field	Description
31-22	Reserved		
21-20	Lane 0	src_l0	Source operand for processing lane 0, 0: operand A; 1: operand B
19-17		opcode_l0	Operation code (Table B.3) for data pre-processing in lane 0
16-15	Lane 1	src_l1	Source operand for processing lane 1, 0: operand A; 1: operand B
14-12		opcode_l1	Operation code (Table B.3) for data pre-processing in lane 1
11-10	Lane 2	src_l2	Source operand for processing lane 2, 0: operand A; 1: operand B
9-7		opcode_l2	Operation code (Table B.3) for data pre-processing in lane 2
6-5	Lane 3	src_l3	Source operand for processing lane 3, 0: operand A; 1: operand B
4-2		opcode_l3	Operation code (Table B.3) for data pre-processing in lane 3
1-0	Mask	mask_src	Source operand for data masking, 0: operand A; 1: operand B

Table B.3 Operation code for matrix data pre-processing

Opcode	Function	Operand
000	None	—
001	Negation	The real part of the complex-valued input
010		The imaginary part of the complex-valued input
011		Both real and imaginary part of the complex-valued input
100	Reserved	
101	Absolute	The real part of the complex-valued input
110		The imaginary part of the complex-valued input
111		Both real and imaginary part of the complex-valued input

Table B.4 Micro-code set for pre-processing cell II (vector data)

Bit	Division	Field	Description
31-14	Reserved		
13	Permutation	perm_en	Permutation enable
12-11		imm_l0	Immediate value input for permutation sequence, processing lane 0
10-9		imm_l1	Immediate value input for permutation sequence, processing lane 1
8-7		imm_l2	Immediate value input for permutation sequence, processing lane 2
6-5		imm_l3	Immediate value input for permutation sequence, processing lane 3
4	Swap	swap_en	Enable for swapping the real and imaginary part of each data operand
3-1	Pre-process	opcode	Operation code (Table B.3) for vector data pre-processing
0	Mask	mask_en	Vector data mask enable

Table B.5 Micro-code set for core-processing cell

Bit	Division	Field	Description
31-17	Reserved		
16-15	Shuffle	opcode_shuf	Operation code (Table B.9) for input data shuffling
14		swap_en_opa	Enable for swapping the real and imaginary part of matrix operand A
13		swap_en_opb	Enable for swapping the real and imaginary part of matrix operand B
12	SIMD	mul_en	Enable for complex-valued multiplication
11		mul_sign	Signed/unsigned multiplication, 0: signed; 1: unsigned
10		add_en_l1	Enable for addition, level-1 adders
9		add_en_l2	Enable for addition, level-2 adders
8		add_sign_l1	Signed/unsigned addition, level-1 adders, 0: signed; 1: unsigned
7		add_sign_l2	Signed/unsigned addition, level-2 adders
6		add_sub_l1a	Addition/subtraction selection, level-1 adder A, 0: addition; 1: subtraction
5		add_sub_l1b	Addition/subtraction selection, level-1 adder B
4		add_sub_l2a	Addition/subtraction selection, level-2 adder A
3		add_sub_l2b	Addition/subtraction selection, level-2 adder B
2-1		opcode_v	Operation code (Table B.10) for vector data operand
0		v_duplicate	Duplication of vector data operand, 0: column-wise; 1: row-wise

Table B.6 Micro-code set for post-processing cell I

Bit	Division	Field	Description
31-18	Reserved		
17	Accumulation	acc_en_l3	Enable for data accumulation, processing lane 3
16		acc_en_l2	Enable for data accumulation, processing lane 2
15		acc_en_l1	Enable for data accumulation, processing lane 1
14		acc_en_l0	Enable for data accumulation, processing lane 0
13		acc_init_l3	Register initialization, processing lane 3, 0: init.; 1: acc.
12		acc_init_l2	Register initialization, processing lane 2
11		acc_init_l1	Register initialization, processing lane 1
10		acc_init_l0	Register initialization, processing lane 0
9	Summation	mux_sum_in	Input multiplexing for lane 3 & 2, 0: straight; 1: swapped.
8		mux_sum_in	Input multiplexing for lane 1 & 0
7		mux_sum_out	Output multiplexing for lane 3 & 2, 0: acc.; 1: sum
6		mux_sum_out	Output multiplexing for lane 1 & 0
5-4	Barrel shifting	bs_opcode	Operation code (Table B.11) for barrel shifting
3-0		bs_imm	Shifting bit count

Table B.7 Micro-code set for post-processing cell II

Bit	Division	Field	Description
31-14	Reserved		
13	Sorting	sort_en	Enable for vector data sorting
12		sort_order	Sorting order, 0: ascending; 1: descending
11		sort_sign	Signed/unsigned sorting, 0: signed; 1: unsigned
10-9	Permutation	perm_src	Permutation sequence input, control code in Table B.12
8-7		perm_imm_l0	Immediate value input for permutation sequence, processing lane 0
6-5		perm_imm_l1	Immediate value input for permutation sequence, processing lane 1
4-3		perm_imm_l2	Immediate value input for permutation sequence, processing lane 2
2-1		perm_imm_l3	Immediate value input for permutation sequence, processing lane 3
0	Mask	mask_en	Enable for vector data mask

Table B.8 Micro-code set for the write-back stage of register bank

Bit	Division	Field	Description
31-19	Reserved		
18	Matrix data writing	reg_m_wen	Enable for matrix data writing
17		reg_m_src	Data source for register writing, 0: matrix bus; 1: vector bus
16		reg_m_vdup	Duplication of vector data output, 0: row-wise; 1: column-wise
15		reg_m_dst	Destination for matrix data writing, 0: GPR; 1: I/O
14-13		reg_m_idx[a]	I/O port number/GPR index for matrix data writing
12	Vector data writing	reg_v_wen	Enable for vector data writing
11		reg_v_dst	Destination for register writing, 0: GPR; 1: I/O
10-7		reg_v_idx	I/O port number/GPR index for vector data writing
6	VPR/	reg_s_wen	Register write enable
5	MMR	reg_s_dst	Destination for data writing, 0: VPR; 1: MMR
4		reg_s_src	Data source for register writing, 0: sorted data; 1: vector data
3-0		reg_s_idx	Register index

[a]"Multiple-of-four" addressing (2 LSBs are discarded) for writing matrix data to GPR, e.g., accessing $0, $4, $8, etc.

Table B.9 Operation code
for input data shuffling in the
SIMD core

Opcode	Operation
00	Complex-valued arithmetic
01	Real-valued arithmetic
10	Complex- & real-valued square operation
11	Reserved

Table B.10 Operation code
for vector data operand in the
SIMD core

Opcode	Operation
00	None
01	Constant multiplication
10	Constant addition
11	Reserved

Table B.11 Operation code
for barrel shifter

Opcode	Operation
00	None
01	Arithmetic shift right (ASR)
10	Logical shift left (LSL)
11	Logical shift right (LSR)

Table B.12 Data source for permutation sequence

Opcode	Operation
00	No permutation
01	Using sorting output as a permutation sequence
10	Using sequence loaded from VPR
11	Using an immediate value input as a permutation sequence

Table B.13 Instruction set for sequencer

Instruction	31-29	28	27-24	23-20	19-16	15-12	11-8	7-4	3-0
Instruction	Opcode	Field							
NOP	000	L[a]	—						
Normal[b]	001	L	offset_1	offset_2	offset_3	offset_4	offset_5	offset_6	offset_7
Loop push[c]	010	—	Loop count			—			
Base config.[d]	011	—	base_1	base_2	base_3	base_4	base_5	base_6	base_7
End of program	100	—	—						ID
Reserved	101	—							
	110	—							
	111	—							

[a] "End-of-loop" flag
[b] Normal instruction, controls the address ("address = base + offset") of distributed configuration memories
[c] Pushing a loop into the stack of the inner loop controller, including link address and loop count
[d] Base address configuration for the distributed configuration memories